SURVIVAL GUIDE FOR SCIENTISTS

Survival Guide for Scientists

Writing – Presentation – Email

Ad Lagendijk

Amsterdam University Press

Second printing 2008

Cover design: Studio Jan de Boer, Amsterdam
Lay-out: ProGrafici, Goes

ISBN 978 90 5356 512 4
e-ISBN 978 90 4850 625 5
NUR 810

To Truusje, Wouter, Guido and Kristel

SUMMARY CONTENTS

Preface		31
Writing Guide for Scientists		
1	Introduction	34
2	Manuscript handling	41
3	Text formatter	50
4	Text structure	55
5	Text content	58
6	Text spelling	75
7	Math	77
8	Figures	82
9	Tables	90
10	Submission	92
11	Referee reports	95
12	Administration	103
13	Reaching out	106
14	Alternative publishing	107
15	Protecting your papers	111
16	About	114
Presentation Guide for Scientists		
1	General	119
2	Process of presentation	126
3	Spoken text	133
4	Slides	146
5	Technical aspects	179
6	Ten commandments	187
7	Checklist	188
8	About	190

Email Guide for Scientists

1	Introduction	193
2	General principles	197
3	Receiving emails	204
4	Sending emails	212
5	Managing account(s)	229
6	Archiving emails	237
7	Security	242
8	Internet Protocols	247
9	About	249

Index 251

DETAILED CONTENTS

Preface 31

WRITING GUIDE FOR SCIENTISTS

1 Introduction 34
- 1.A Target group 34
- 1.B Goal of your paper 35
  - 1.B.1 Help your readers 35
  - 1.B.2 Profile of your readers 35
  - 1.B.3 Quality of your papers 35
- 1.C Discussion groups 35
- 1.D General advice for authors 36
  - 1.D.1 Help your coauthors 36
- 1.E Format of the *Writing Guide* 36
  - 1.E.1 Publication form 36
    - 1.E.1.A Printed version 36
    - 1.E.1.B Digital version 36
  - 1.E.2 Navigation 37
  - 1.E.3 Size of the *Writing Guide* 37
  - 1.E.4 Text formatter 37
- 1.F How to use the *Writing Guide*? 37
  - 1.F.1 Required prior knowledge 37
  - 1.F.2 Further study 37
    - 1.F.2.A Courses in writing scientific papers 38
- 1.G Improving your English 38
- 1.H Culture 39
- 1.I Limitations 39
- 1.J Male chauvinism 39
- 1.K Ethics 39
- 1.L Conventions in the *Writing Guide* 39
  - 1.L.1 Double quotes 40
  - 1.L.2 Single quotes 40
- 1.M Commercial products 40

2 Manuscript handling 41
2.A Consistency 41
2.B Manuscript types 41
2.B.1 Generic scientific texts 41
2.B.1.A Extreme size constraint 42
2.B.1.B Theses 42
2.B.1.B.1 Consistency 42
2.B.2 Some special scientific texts 42
2.B.2.A Computer programs 42
2.B.2.A.1 Manual 43
2.B.2.B Comments 43
2.B.2.C Conference abstracts 43
2.B.2.C.1 Style of conference abstracts 43
2.B.2.C.2 Credit in conference abstracts 44
2.B.2.D Referee reports 44
2.B.2.E Grant proposals 44
2.B.2.E.1 Formatter 44
2.B.3 Your own layout 44
2.B.3.A Digital format 44
2.C Where to publish? 45
2.D Corrections by coauthors 45
2.D.1 Bilateral correction process 45
2.D.2 Checking coauthors 46
2.E Corrections by one coauthor 46
2.E.1 White space in concepts 46
2.E.1.A Communication with first author 47
2.E.1.B Size limit 47
2.E.1.C Final format 47
2.E.2 Version control 47
2.E.2.A Long papers 47
2.E.2.B Meeting of authors 47
2.E.2.C Dead line for active corrector 48
2.E.2.D Incremental texts 48
2.E.2.E Duplication problems 48

3 Text formatter 50
3.A *Tex* family 50
3.A.1 Record: length and ending of record 50
3.A.2 Compatibility 51
3.A.2.A Converting from old versions 51
3.A.3 Other text formatters like *MS-Word* 51
3.B *MS-Word* survival 51
3.B.1 *Word 2007* 51
3.B.2 Justification 52
3.B.3 Left alignment 52

3.B.4 Full justification 52
3.B.5 Mathematical equations 52
3.B.6 Recovery 52
3.B.7 Navigation 53
3.B.8 Dictionaries 53
3.B.9 Macros 53
3.B.10 Templates 53
3.B.11 Version control 54
3.B.12 Wid ows 54

4 Text structure 55
4.A Organization of content 55
4.A.1 Margins 55
4.A.2 Justification 55
4.A.3 Subheadings 56
4.A.3.A Hidden subheadings 56
4.A.4 Paragraphs 56
4.A.5 Spaghetti text 56
4.A.6 Parking places 57
4.A.7 Standard partitioning 57
4.A.8 Smooth text transitions 57
4.B Length of sentences 57

5 Text content 58
5.A Obligatory items 58
5.A.1 Title 58
5.A.2 Authors 58
5.A.2.A First author 58
5.A.2.A.1 Many first authors 58
5.A.2.B Coauthors 59
5.A.2.B.1 Placeholder 59
5.A.2.B.2 Individual contributions 59
5.A.2.B.3 Contribution of world specialist 59
5.A.2.C Order of authors 59
5.A.2.D Corresponding author 60
5.A.2.E Spelling of names 60
5.A.2.E.1 Use of first names in author list 60
5.A.3 Affiliations 60
5.A.3.A Change of affiliation 61
5.A.4 Abstract 61
5.A.4.A Keywords 61
5.A.4.B Subject classification 61
5.A.5 Date 62
5.A.6 Introduction 62
5.A.7 Conclusion 62

5.A.8 Acknowledgement 63
5.A.8.A People in acknowledgement 63
5.A.8.B Refused coauthors 63
5.A.8.C Institutions in acknowledgement 63
5.A.9 List of references 63
5.A.9.A Choice of references 64
5.A.9.B Blockbuster 64
5.A.9.C Reviewers 64
5.A.9.D Own work 64
5.A.9.E Unpublished 65
5.A.9.E.1 Internet 65
5.A.9.F Preprint server 65
5.A.9.G Accepted for publication 65
5.A.9.H Private communication 66
5.A.9.I Non-journal references 66
5.A.9.J Foreign language 66
5.A.9.K Spelling of names 66
5.A.9.L Use of "et al." 66
5.A.9.M Use of "ibid." 67
5.A.10 List of figure captions 67
5.B Experiment 67
5.B.1 Arbitrary units 68
5.C Theory 68
5.C.1 Arbitrary units 68
5.C.2 Proportionality sign 68
5.C.3 Simulations 69
5.D Theory and experiment 69
5.D.1 Your new model 69
5.D.2 Message to theoreticians 69
5.D.3 Message to experimentalist 70
5.D.3.A Scaled units 70
5.E Priority 70
5.E.1 Credit to others 71
5.E.1.A Criticizing prior work 71
5.E.2 First-time claim 71
5.E.3 Emphasize own work 72
5.F Grammar 72
5.F.1 Use of "This means" and "That implies" 72
5.F.2 When "that" or "which" leads a clause 72
5.F.3 Same words over and over again 72
5.F.3.A Thesaurus 73
5.F.3.A.1 Online 73
5.F.4 Absolute statements 73
5.F.5 Exclamation marks 73
5.F.6 Emphasized and underlined text 73

5.F.7 Active or passive form 73
5.F.7.A Abstract subjects 74

6 Text spelling 75
6.A Consistency 75
6.B Hyphens 75
6.C Quotation marks 75
6.D Full capital abbreviations 75
6.E Conventional abbreviations 76
6.F Non-alphabetic characters 76

7 Math 77
7.A Conventions 77
7.A.1 Alphabet 77
7.A.1.A Vectors 77
7.A.1.B Diacritics 78
7.A.1.B.1 Abbreviations as subscript 78
7.A.2 Naming conventions 78
7.A.3 Units and constants 78
7.A.3.A International bodies 78
7.A.3.B Dimensions 78
7.B Display math (in contrast to inline math) 78
7.B.1 Punctuation marks 79
7.B.2 Equation signs and definition sign 79
7.B.3 Numbering of equations 79
7.B.4 Symbols in equations 80
7.B.5 Brackets 80
7.B.6 Use of appendices 80
7.B.7 Abuse of same symbol 80
7.C Inline math 80
7.C.1 Integration with text 80
7.C.2 Beginning of sentence 80
7.C.3 Headings with math 81

8 Figures 82
8.A Default setting 82
8.B Text aspects of figures 82
8.B.1 Manuscript text referring to figures 82
8.B.2 Figure captions 83
8.B.2.A Example of bad caption 83
8.B.3 Font size in figures 83
8.B.4 Fonts in figure text (labeling of axes etc.) 83
8.C Postscript 84
8.C.1 Compatibility 84
8.C.2 Bounding box 84

8.D Non-postscript 85
8.E Drawing elements 85
8.E.1 Frames 85
8.E.2 Axes 85
8.E.2.A Two different X or Y axes 85
8.E.2.B Tick marks 85
8.E.2.C Labels 85
8.E.3 Grid lines 86
8.E.4 Line thickness 86
8.E.5 Line style 86
8.E.6 Symbols 86
8.E.7 Busy figures 86
8.E.8 Color 86
8.E.8.A Grey scales 87
8.E.9 Kill dangerous colors 87
8.F Colorful figures 87
8.F.1 Three-dimensional figures 87
8.F.1.A Professional lighting 88
8.F.1.B Interpretation 88
8.F.2 Color coding 88
8.F.3 Cross-sections 88

9 Tables 90
9.A Columns and rows 90
9.B Justification 90
9.C Caption and title 90
9.D Reuse 91

10 Submission 92
10.A Permission 92
10.B Submission letter 92
10.B.1 *Nature* and *Science* 93
10.B.1.A Manuscript 93
10.B.1.A.1 Good-news show 93
10.B.1.A.2 Math 94
10.B.1.A.3 Graphics 94
10.B.1.B Submission letter 94
10.C Mode of transport 94

11 Referee reports 95
11.A Answering reports 95
11.A.1 Who will write the replies? 95
11.A.2 Styled text 95
11.A.3 Style of rebuttal 96
11.A.4 Header of replies 96

11.A.5 Opening remarks rebuttal 96
11.A.6 Scientific conflicts 97
11.A.6.A Referee finds a flaw 97
11.A.7 Length of rebuttal 97
11.A.8 Manuscript modifications 97
11.A.8.A One referee at a time 97
11.A.8.B Modifications due to other referee 98
11.A.9 Conflicting referee reports 98
11.A.10 Additional references98
11.B Perseverance 98
11.C Letter to the editor 98
11.D Resubmission package 99
11.E Failed submission 99
11.E.1 Protest against refusal of review 100
11.F Writing referee reports 100
11.F.1 Pride 100
11.F.2 Reviewing load 101
11.F.3 Alternative referees 101
11.F.4 Standard format 101
11.F.5 Quality of manuscript 102

12 Administration 103
12.A Publication costs 103
12.B Reprint orders 103
12.C Mailing list 103
12.D Digital Archive 104
12.D.1 Group file server 104
12.D.2 Saving files 104

13 Reaching out 106
13.A Publicizing your work 106

14 Alternative publishing 107
14.A Official: preprint server 107
14.B Unofficial publishing 107
14.B.1 Format 107
14.B.1.A Pdf files 107
14.B.1.A.1 Pdf fonts 108
14.B.1.A.2 Layout of paper 108
14.B.1.A.3 Figures in pdf files 108
14.B.1.A.4 Figures in single-column pdf files 109
14.B.1.A.5 Figures in two-column pdf files 109
14.B.1.A.6 Navigation 109
14.B.1.B Postscript 109
14.B.1.C Web page 109

14.B.1.C.1 Html formatting 110
14.B.1.C.2 Navigation 110
14.B.1.D Desktop publishing 110
14.B.2 Copy right and web posting 110

15 Protecting your papers 111
15.A Reasons 111
15.A.1 Commercial reasons 111
15.A.2 First discovery 111
15.B Protection solutions 111
15.B.1 Regular publications 111
15.B.2 Watermarks 111
15.B.3 *MS-Word* documents 112
15.B.4 Protecting pdf files 112
15.B.4.A Commercial copy protection 112
15.B.4.B Access protection 112
15.B.4.C Disable text copying once and for all 112
15.B.5 Websites 113

16 About 114
16.A Abbreviations 114
16.B Trademarks 114

PRESENTATION GUIDE FOR SCIENTISTS

1 General 119
1.A Target group 119
1.B Goal of presentations in general 120
1.B.1 Classification of talks 120
1.B.2 Talks of type a) 120
1.B.3 Talks of type b) 120
1.B.4 Talks of type d) 120
1.C Goal of the *Presentation Guide*
121
1.C.1 Discussion groups 121
1.D Presentation formatter 121
1.D.1 Are hints limited to users of *PowerPoint*? 122
1.E Format of the *Presentation Guide* 122
1.F Publication form of the *Presentation Guide* 122
1.F.1 Publication form 122
1.F.1.A Printed version 122
1.F.1.B Digital version 122
1.F.2 Navigation 122
1.G Use of the *Presentation Guide* 122

1.G.1 Communities 123
1.H Size of the *Presentation Guide* 123
1.I Male chauvinism 123
1.J Training courses 123
1.J.1 Improving your English 124
1.K Conventions of the *Presentation Guide* 124
1.K.1 Double quotes 124
1.K.2 Single quotes 124
1.K.3 Vague qualifiers 124
1.K.4 Commandments 124
1.K.5 To-Do list for you, readers of this Guide 125
1.L Legal disclaimer 125
1.M Slick presentations 125
1.N Preparation time 125

2 Process of presentation 126
2.A Conferences 126
2.A.1 Parallel 126
2.A.2 Plenary 126
2.A.3 Preference schedule for your talk 127
2.A.4 Leaving and entering audience 127
2.A.5 Position in auditorium 127
2.A.6 Hotel 127
2.A.6.A Room mates 128
2.B Single event 128
2.B.1 Group meeting 128
2.C Evaluation committees 129
2.C.1 Crucial prior knowledge 129
2.C.2 Acquaintance 129
2.C.3 Handouts 129
2.D Attitude 130
2.D.1 Body movements 130
2.D.2 Body position 130
2.D.3 Dressing 131
2.E Handouts 131
2.F Rehearse 131
2.F.1 Without audience 131
2.F.2 Important talk 131
2.F.2.A Reviewers 132
2.F.2.B Procedure 132
2.F.2.C Without audience 132

3 Spoken text 133
3.A Allotted time 133
3.A.1 Defend your time 133

3.B	Volume, tone and pace			133
	3.B.1 Volume			133
	3.B.2 Pace			134
	3.B.3 Female speakers			134
	3.B.4 Nervousness			134
		3.B.4.A Speech support		134
			3.B.4.A.1 Speech therapists	135
3.C	Language			135
	3.C.1 Stopgaps			135
	3.C.2 Synonyms			135
3.D	Scientific level			135
	3.D.1 Seminar			136
	3.D.2 Colloquium			136
	3.D.3 Plenary			136
	3.D.4 Popular talks			136
3.E	Too much information (TMI)			136
3.F	Funnel			137
3.G	Tell one story			137
	3.G.1 Operators			137
	3.G.2 No distraction			137
3.H	Social behavior			138
	3.H.1 Credit			138
		3.H.1.A Opening remarks		138
		3.H.1.B Closing remarks		138
		3.H.1.C Conference		138
		3.H.1.D Colloquium		138
		3.H.1.E Your ego		139
		3.H.1.F Lack of credit		139
3.I	Questions and interference			139
	3.I.1 Repeat the question			139
		3.I.1.A Advantages		140
			3.I.1.A.1 Audience participates	140
			3.I.1.A.2 Gain of time	140
			3.I.1.A.3 Reduce effect of question	140
	3.I.2 Politeness			140
		3.I.2.A Crackpots		140
	3.I.3 During talk			140
		3.I.3.A Test of audience		141
		3.I.3.B Disturbing expert		141
		3.I.3.C Too many questions		141
	3.I.4 At end			141
		3.I.4.A Comments		141
	3.I.5 Hostile participants			142
		3.I.5.A Priority claim		142
		3.I.5.B Trap		142

3.I.5.C Backup slides 142
3.I.5.D Humor 142
3.I.6 Hostile audience 143
3.I.7 Evaluation committee 143
3.I.7.A Atmosphere 143
3.I.7.B Hostile questions 143
3.I.7.C Divide et impera 143
3.J Requests from audience 144
3.J.1 Unpublished material 144
3.J.2 Life internet connections 144
3.K Slide by slide 144
3.K.1 Slide contains plots 145

4 Slides 146
4.A General structure 146
4.A.1 File size 146
4.A.2 Reusability 146
4.A.3 Composition 147
4.A.3.A Landscape 147
4.A.3.B Separators 147
4.A.3.C Plots 147
4.A.3.D Alignment 148
4.A.3.E Margins 148
4.A.3.E.1 Top 148
4.A.3.E.2 Bottom 148
4.A.3.E.3 Left and right 148
4.A.3.E.4 Large left margin 148
4.A.3.F Guides 148
4.A.4 Credit 149
4.A.4.A Full references 149
4.A.4.B Advertising own work 149
4.A.4.B.1 Website 150
4.A.4.B.2 Hyperlinks 150
4.B Contrast and colors 150
4.B.1 Object colors 150
4.B.1.A Colored background 150
4.B.1.B Graded background 151
4.B.1.C Background texture 151
4.B.1.D Background picture 151
4.B.2 Slide background 151
4.B.2.A Acceptable background 152
4.B.3 Dangerous colors 152
4.B.3.A Default colors 152
4.B.4 Shadowing 152
4.C Text properties 153

4.C.1 Font 153
4.C.1.A Font size 153
4.C.1.B Bold 153
4.C.1.B.1 Serif 153
4.C.1.B.2 Non-serif 153
4.C.1.C Underlining 154
4.C.1.D Font color 154
4.C.1.E Collection of fonts 154
4.C.2 Spelling and grammar 154
4.C.2.A Complete sentences 154
4.C.2.B Periods 154
4.C.2.C Capitals 154
4.C.2.D Definite articles 155
4.C.2.E Exclamation sign 155
4.C.2.F Question mark 155
4.C.2.G Space before unit 155
4.C.3 Alignment 155
4.C.3.A Align left 155
4.C.3.B Align right if text connects to right 155
4.C.3.C Centered alignment 155
4.C.3.D Multi-lined texts 155
4.C.4 Title 156
4.C.4.A Frames 156
4.C.5 Text frames 156
4.C.5.A Good text frames 156
4.D Outlined text (bullets) 157
4.D.1 Header 157
4.D.2 Type of bullets 157
4.D.2.A Balls as character 157
4.D.2.B Numbered outline 158
4.D.2.C Other bullet characters 158
4.D.2.D Bullets as pictures 158
4.D.2.E Color coding 158
4.D.2.F Formatting outline lines 158
4.D.2.F.1 Font 158
4.D.2.F.2 Line spacing 158
4.D.2.F.3 Multiple lines in one outline item 158
4.D.2.F.4 Horizontal alignment 159
4.D.2.F.5 Vertical alignment 159
4.D.2.F.6 Multiple-line spacing 159
4.D.2.F.7 Professional examples 159
4.E Tables 160
4.E.1 Alignment 160
4.E.2 Absolute numbers 160
4.E.3 Error bars 160

4.F Math 160
4.F.1 Math from copy/paste 160
4.F.1.A Contrast 161
4.F.1.B Math with *Tex* 161
4.F.1.C Math with *MathType* 161
4.F.2 Numbering equations 162
4.F.3 Size 162
4.F.4 Style 162
4.F.4.A Inline math 162
4.F.4.A.1 Fonts 162
4.F.4.B Font size 162
4.F.4.C Coloring equations 162
4.F.4.D Boxing equations 163
4.G Graphics 163
4.G.1 Figures not being plots 163
4.G.1.A Bitmap or vector 163
4.G.1.B Bitmaps with poor contrast 164
4.G.1.C Gif transparent 164
4.G.1.D Collection of pictures 164
4.G.1.E Animation 164
4.G.2 Scientific plots 164
4.G.2.A Rule from the *Writing Guide* 164
4.G.2.B Number of figures 164
4.G.2.C Second X-axis and Y-axis 165
4.G.2.D Frames around figures 165
4.G.2.E Captions and titles 165
4.G.2.F Archived plots 165
4.G.2.G Cleaning up plots 165
4.G.2.H Repair figures 166
4.H Obligatory slides 166
4.H.1 Master 166
4.H.1.A Slide titles 166
4.H.1.B *PowerPoint* templates 167
4.H.1.C Frames and logos 167
4.H.1.D Dates and occasion 167
4.H.1.E Progress macros 167
4.H.1.F Numbering 168
4.H.1.G Bottom margin 168
4.H.2 Front slide 168
4.H.3 Title slide 168
4.H.4 Contents slide 168
4.H.5 Introduction slide 168
4.H.6 Progress 169
4.H.6.A Progress slide 169
4.H.6.B Slide numbering 169
4.H.6.C Headers and footers 170

4.H.7 Coworkers 170
4.H.7.A Science agencies 170
4.H.8 Conclusions 170
4.H.8.A Style 170
4.H.8.B Combining with other information 171
4.H.8.C One-liners 171
4.H.8.D Size 171
4.H.8.E New information 171
4.H.8.F Following slides 171
4.H.8.G Afterburner 171
4.I Animation 171
4.I.1 Animated text 172
4.I.2 Animated gif 172
4.I.3 Java applet 172
4.I.4 Video 173
4.I.4.A *Windows Media Player* 173
4.I.5 Extra files 173
4.I.5.A Too many or too slick movies 173
4.J Navigation through slides 173
4.J.1 Presenter view 174
4.K Transition 174
4.L Software incompatibilities 174
4.L.1 *PowerPoint* 174
4.L.1.A Backward compatibility 175
4.L.1.B Auto recovery 175
4.L.1.C Improved auto-recovery 176
4.L.2 Proprietary fonts 176
4.L.3 Missing fonts 176
4.L.3.A Embedding of fonts 177
4.L.3.A.1 Pack and Go (Obsolete now) 177
4.L.3.B Old *PowerPoint* versions 177
4.L.3.B.1 Installation advice 177
4.L.3.B.2 Font-saving problems 178
4.L.4 Printing problems with handouts 178
4.L.5 *PowerPoint* 2007 178

5 Technical aspects 179
5.A At home 179
5.A.1 Video projector (obsolete now) 179
5.A.2 Laptop 179
5.A.2.A Power management 179
5.A.2.B Laptop rebooting 179
5.A.2.C Laptop resolution 180
5.A.2.D Wifi 180
5.A.2.E Wireless mouse 180

5.A.2.F Bluetooth 180
5.A.3 Accessories 181
5.A.3.A Adapters for wall plugs 181
5.A.3.B Plugs and sockets 181
5.A.3.C Ethernet 181
5.A.3.D Phone and modem 181
5.A.3.D.1 Fixed-line phone 181
5.A.3.D.2 Mobile phone 181
5.A.3.E Pointer 182
5.A.3.F Count-down timer 182
5.A.3.G Flashlight 182
5.A.4 Print handout 182
5.A.5 Backup 183
5.A.5.A For use at conference 183
5.A.5.B For reuse 183
5.A.6 Personal care 183
5.B Traveling 183
5.B.1 Airport 183
5.B.1.A Wireless 184
5.B.2 At conference 184
5.B.2.A Your own laptop 184
5.B.2.A.1 When to switch on laptop 184
5.B.2.A.2 Start *PowerPoint* 184
5.B.2.B Troubleshooting 184
5.B.2.B.1 Be the last to test video projector 185
5.B.2.C Lighting 185
5.B.2.D Fixed position 185
5.B.3 Conference laptop 185
5.B.4 Sound 186
5.B.5 Boards 186
5.B.6 Flip-charts 186
5.B.7 After care 186
5.B.7.A At home 186

6 Ten commandments 187

7 Checklist 188
7.A The night before you leave 188

8 About 190
8.A Trademarks 190

EMAIL GUIDE FOR SCIENTISTS

1 Introduction 193
- 1.A Goal of the *Email Guide* 193
  - 1.A.1 Discussion groups 193
- 1.B Target group 194
  - 1.B.1 Technical text 194
  - 1.B.2 Male chauvinism 194
- 1.C Format of the *Email Guide* 194
  - 1.C.1 Size of the *Email Guide* 194
  - 1.C.2 Publication form 195
    - 1.C.2.A Printed version 195
    - 1.C.2.B Digital version 195
- 1.D How to use the *Email Guide* 195
  - 1.D.1 Navigation 195
- 1.E Conventions 195
  - 1.E.1 Double quotes 195
  - 1.E.2 Single quotes 195
  - 1.E.3 Hyperlinks 196
- 1.F Legal disclaimer 196

2 General principles 197
- 2.A Face to face or by email 197
- 2.B Confirm by email 197
- 2.C Educate the world 198
  - 2.C.1 Colleagues 198
  - 2.C.2 Educate secretaries 198
  - 2.C.3 Educating managers 198
  - 2.C.4 Educating your students 198
- 2.D Internet philosophy 198
  - 2.D.1 *Requests for Comments* 199
  - 2.D.2 ASCII as standard 199
- 2.E Email 'programs' 199
  - 2.E.1 Server or client 199
  - 2.E.2 Email servers 200
  - 2.E.3 Email clients 200
    - 2.E.3.A Webmail 200
      - 2.E.3.A.1 Clumsy web interface 201
      - 2.E.3.A.2 Webmail as a last resort 201
      - 2.E.3.A.3 Copy of recent emails 201
  - 2.E.4 *Windows* or *Unix* 201
- 2.F To-Do list 202
- 2.G Raw email address 202
  - 2.G.1 Hiding email address 202
- 2.H Your own email address 203

3 Receiving emails 204
3.A Filtering 204
3.B Spam 204
3.B.1 Commercial anti-spam software 205
3.B.2 Included hyperlinks 205
3.B.3 Non-commercial spam 205
3.B.3.A Get even with offenders 205
3.B.4 Spam reporting 206
3.C Viewing 206
3.C.1 Adding columns 206
3.C.2 Font 207
3.C.3 Format 207
3.C.4 Mail check frequency
207
3.D Acting on received emails 207
3.D.1 No action required 207
3.D.2 Single action required 208
3.D.2.A No more Mr. Nice Guy 208
3.D.2.B Action is writing email(s) 208
3.D.2.C Deadlines 208
3.D.2.C.1 No deadline 208
3.D.2.C.2 Unreasonable deadlines 208
3.D.2.C.3 Extremely unreasonable requests 209
3.D.3 Multiple actions required 209
3.D.3.A Action for more people 209
3.D.3.B Response forms 210
3.D.3.C Endless actions 210
3.E Dangerous attachments 210
3.E.1 Dangerous attachments 210
3.F Dangerous emails 211

4 Sending emails 212
4.A There is no unsend 212
4.B Formatting of your message 212
4.B.1 ASCII 212
4.B.1.A ASCII styling 212
4.B.2 Rtf 213
4.B.3 Html 213
4.B.3.A Figures 213
4.B.3.B Lack of standardization 213
4.B.4 Hyperlinks 214
4.B.4.A Wrapping 214
4.B.5 Hard returns 214
4.B.6 Bad characters 215
4.B.7 Font 215
4.C Requested action 215

4.C.1 Single request only 216
4.C.2 Requested response time 216
4.D Special emails 216
4.D.1 Sensitive emails 216
4.D.2 Formal emails 216
4.D.2.A Letter simulation 217
4.D.2.B Name and address information 217
4.D.2.C Reference 217
4.D.2.D Salutation 217
4.D.2.E Recap 217
4.D.2.F Data 217
4.D.2.G Attachments 217
4.D.2.H Formal cc 218
4.D.2.I Formal bcc 218
4.D.2.J Example 218
4.D.2.K Fax has legal consequences 218
4.D.3 Confidential emails 219
4.E Header 219
4.E.1 Automatic address 219
4.E.2 Reuse of email addresses 219
4.E.2.A Address books 219
4.E.3 Header part *To* 219
4.E.3.A Email to all 220
4.E.4 Header part *Cc* 220
4.E.5 Header part *Bcc* 220
4.E.6 Subject 220
4.E.7 Attachments 221
4.E.7.A Reason for attachment 221
4.E.7.B Stupid attachments 221
4.E.7.C Color 221
4.E.7.D Naming of file 221
4.E.7.E Business card 221
4.F Body 222
4.F.1 Style 222
4.F.2 Salutation 222
4.F.2.A Friends 222
4.F.2.B No salutation 223
4.F.3 Requested action 223
4.F.4 Trivial body 223
4.F.5 Signature files 223
4.F.6 Grammar and spelling 223
4.F.6.A Date 224
4.F.6.A.1 Relative dates 224
4.F.6.B Time 224
4.F.7 Figures 224

4.F.8 Tables 224
4.F.9 Planning a group event 224
4.G Replying to emails 225
4.G.1 Check the end of your message 225
4.G.2 Send your own format 226
4.G.3 Standard reply 226
4.G.4 Trimming 226
4.G.5 Reply to all 227
4.G.6 Waiting time before you reply 227
4.G.7 Out-of-office reply 227
4.H Redirecting 228
4.I Forwarding 228
4.I.1 To one person 228
4.I.2 To more people 228
4.J Credit card 228

5 Managing account(s) 229
5.A Multiple accounts 229
5.A.1 Prestige of account name 229
5.A.2 Automatic forwarding 229
5.A.2.A Hard forward 230
5.A.2.B Soft forward 230
5.A.3 Multiple POP3/SMTP servers 230
5.A.4 *Exchange Server* plus POP3/SMTP 231
5.A.4.A Sending emails 231
5.A.4.B Filtering emails 231
5.A.5 Multiple Exchange Servers 231
5.B Synchronization issues 232
5.B.1 No syncing required 232
5.B.1.A Fully Localized 232
5.B.1.B One portable computer only 232
5.B.1.C Web interface only 232
5.B.1.D Network drive only 233
5.B.2 Syncing required 233
5.B.2.A Slow servers are a fact of life 233
5.B.2.B VPN 233
5.B.2.C Synchronization challenges 233
5.B.3 Movable storage 234
5.B.3.A USB device 234
5.B.3.B Synchronization 234
5.C Retrieving emails 235
5.C.1 Received mail 235
5.C.1.V.1 Inbox copy 235
5.C.2 Reply address 235
5.D Sending email 236

6 Archiving emails 237
6.A Keeping old emails 237
6.B Archiving formats 237
6.B.1 Ideal format 237
6.B.2 Proprietary file formats 238
6.B.3 *Outlook* format 238
6.B.3.A Pst files 238
6.B.3.A.1 Two versions of *Outlook* 238
6.C Filing system 239
6.C.1 Sent emails 239
6.C.1.A.1 Attachments 239
6.C.2 Received emails 239
6.C.2.A Attachments 239
6.C.2.A.1 Saving with attachments 240
6.C.2.A.2 Saving separately 240
6.C.2.A.3 Version control 240
6.D Finding 240
6.E Backing up 241

7 Security 242
7.A ICT group 242
7.A.1 Are you a computer professional? 242
7.B Encryption 242
7.C Privacy 243
7.C.1 Remaining anonymous 243
7.C.1.A Sending anonymous emails 243
7.C.2 Receiving emails anonymously 243
7.D Protecting your data 243
7.D.1 Office PC 243
7.D.1.A Protect partitions 244
7.D.1.B No traces 244
7.D.1.C USB disk 245
7.D.2 Network drive 245
7.D.2.A Encryption 245

8 Internet Protocols 247
8.A Uri's 247
8.B Name servers 247
8.B.1. Spoofing 247
8.C Core protocols 248
8.D Ports 248
8.E Email RFC 248

9 About 249
9.A Abbreviations 249
9.B Trademarks 249

INDEX 251

PREFACE

Communication is part of almost any professional activity. For researchers, writing scientific papers and giving scientific presentations is a daily ritual. Scientists that do not practice these communication activities will not make any contribution to science. Fortunately, communication skills can easily be learnt by any professional. Unfortunately, many scientists – due to a chronic lack of self-knowledge – do not feel any necessity for improving their own communication competences. As a result, scientists often (50% of the time) give bad talks and often (50% of the cases) write low-quality manuscripts. In addition, many courses are given by people whose profession it is to give those courses. Lessons given by non-active researchers are hardly ever useful, but invariably a waste of time.

In the present guide I give very practical how-to advice on essential topics such as the foundations for writing scientific texts (including dealing with referees and editors), presenting data and research information, and the writing of collegial, efficient emails. Each section is organized as a collection of short rules, outlined and numbered in a logical order as self-explanatory pieces of information – allowing the reader the freedom to study any number of them in any desired order.

A number of the hints are certainly politically incorrect, but they are all the more useful and can be found nowhere else.

Originally the guides were aimed primarily at undergraduate, graduate, and postdoctoral students in the natural sciences. But I have discovered that also more senior scientists will profit from it.

But it doesn't stop there: many of my hints, in particular those referring to presentations, are of invaluable use for a much broader audience of professionals, up to consultants and (public) managers.

Studying the full text will cost less than one hour and a half. Implementing the hints will immediately result in far better presentations, far better scientific papers and far better emails. In the beginning, adoption of the guidelines might cost somewhat more time. But in the end it will be amply rewarding: Your manuscripts will be more easily accepted by editors and referees, they will be better read, and better cited. Your talks will stand out. And all of this achieved by – in the long run – spending altogether less time on these activities.

This guide represents my views and my advice. But it is also meant to be your guide. Navigate to www.sciencesurvivalblog.com to contribute, discuss or criticize my hints. The whole community will profit. Next editions will have your hints in them as well.

This text has evolved as the result of collaborations with PhD students, postdocs and senior colleagues of the *University of Amsterdam*, *University of Twente* and the *FOM-Institute for Atomic and Molecular Physics*. I am grateful for their input.

I am thankful to Sanford Bingham, CEO of *FileOpen Systems Inc.*, for his generous (software) support in the early stage of this project.

Ad Lagendijk

WRITING GUIDE FOR SCIENTISTS

1 INTRODUCTION

The text you are reading right now, the *Writing Guide for Scientists* (from now on: *Writing Guide*), lays out a set of fundamental rules regarding writing scientific texts.

This "writing" is interpreted in a broad way. The rules include advice on submission strategies, on how to deal with editors and referees, on web posting, and advise on many more aspects of producing scientific articles.

This tutorial in itself is not a scientific text, and therefore the principles presented here hardly apply to the text itself.

The *Writing Guide* is part of the *Survival Guide for Scientists* (from now on the *Survival Guide*). Occasionally there might be general referrals in the *Writing Guide* to other parts of the *Survival Guide*. However, the *Writing Guide* is supposed to be self-contained. For completeness we present here the names of all the guides that together constitute the *Survival Gui*de for Scientists:

- *Presentation Guide for Scientists,* or short: *Presentation Guide*; as an addendum to the *Presentation Guide* we have published the *Example Guide*
- *Email Guide for Scientists,* or short: *Email Guide*
- *Writing Guide for Scientists,* or short: *Writing Guide*
- *Survival Guide for Junior Scientists,* or short: *Junior Guide*
- *Survival Guide for Senior Scientists,* or short: *Senior Guide*

1.A Target group

The target group I had originally in mind was physics undergraduate and graduate students, and physics postdocs. From experience I have discovered that senior physicists could also profit from studying the set of instructions laid out in the *Writing Guide*. (Just check a recent article in an international physics journal to appreciate this argument: quite a number of my rules are violated in these papers.)

The text is very likely also beneficial for mathematicians and for workers in other natural science disciplines, like chemistry and astronomy. The content is highly modular. Researchers in other fields can easily skip parts they consider too closely related to physics, or which they deem irrelevant for other reasons.

The number of cross-links in the guide is kept to a minimum. Each item can be studied on its own.

1.B Goal of your paper

If you do not agree with the goals I will present in the following list, there is no need for you to continue to study the *Wring Guide*.
The primary goals are:
I to maximize the number of readers;
II to minimize the time needed to 'read' your paper;
III to maximize the fraction of satisfied readers;
IV to maximize the number of citations the paper will get.

The world of scientific publications is well characterized by the one-liner: "Get cited, or get lost".

1.B.1 Help your readers

To achieve the above four objectives you must make life easy and pleasant for your readers. If your intention is to make life easy on yourself you should quit physics (or for that matter any other natural science; law school might be more appropriate).

By only diagonally browsing your article, or by just studying its conclusion, or by merely looking at one of your figures, the reader should get valuable information. Your paper should allow for smooth and quick navigation. Readers might already know half of the content of your writing. Assist them in spotting the new material fast and help them skipping the rest.

1.B.2 Profile of your readers

Your average reader is mildly interested. Never expect scientists to go through your article from beginning to end. They will never read your references (but they will certainly check whether or not you cite them).

Professional researchers have to browse many, many papers on a daily basis and are continuously looking for a pretext to put your paper aside. You can call yourself lucky if they grant you ten seconds to obtain a first impression.

1.B.3 Quality of your papers

If you agree with the four principal goals laid out above, you can profit a lot from following the hints of the *Writing Guide*. Only some of my hints (but in my opinion really only a few) are a matter of taste.

Studying the principles will help researchers to write better papers, to produce superior replies to referees, and to become better referees.

1.C Discussion groups

The author of the *Writing Guide*, that is me, has over 30 years of experience of writing scientific papers. In my opinion many of my hints are crucial for writing good texts. But it is also your guide: if you do not agree with one, some, or many of my hints, post your own ideas at our weblog www.sciencesurvivalblog.com. If more people agree with you, the *Writing Guide* will be improved by implementing your advice.

http://www.sciencesurvivalblog.com

1.D General advice for authors

Do not reinvent the wheel. Scientific texts have evolved over hundreds of years. Copy, steal, plagiarize, reuse, or whatever your favorite expression is for the action of using text structures, texts, mathematical equations, and figures, generated by senior and junior colleagues and by competitors. In many cases no source has to be acknowledged.

1.D.1 Help your coauthors

An additional advantage of closely following the advice given in this guide, is that your coauthors, including your supervisor, will be very satisfied with the efficient way you handle the shared papers.

Some of my advice will mean some unexpected hard work on the side of the (first) author (that is you) of a scientific paper. On the long run this will pay off.

1.E Format of the *Writing Guide*

This tutorial is organized as a large collection of short rules, outlined and numbered in a hierarchical way. The directives often represent an independent piece of information, so that the reader can work through any number of items in any desired sequence.

1.E.1 Publication form

The *Writing Guide* is available in basically two forms: as a book and as an ebook.

1.E.1.A Printed version

In the printed version the first three guides (*Presentation Guide, Email Guide,* and *Writing Guide*) are collected in one volume.

1.E.1.B Digital version

The *Writing Guide* will also be available as ebook (protected pdf). The pdf file will be prepared in cooperation with *FileOpen Systems.*

http://fileopen.com

1.E.2 Navigation

Navigation through the digital version is easy: there are bookmarks in the pdf version.

The printed version will be bound in such a way that it can be read hands-free.

1.E.3 Size of the *Writing Guide*

Suggestions for additions, corrections, or other ideas for changes and improvements are welcome if they do not make the text much longer. It is my intention to keep the size of the *Writing Guide* to a maximum of about 600 paragraphs. This size constraint ensures that scientists can read the whole text in less than half an hour. Increasing the size beyond this limit would deter too many members of the target group.

1.E.4 Text formatter

In physics and mathematics the standard text formatters are members of the *Tex* family (*LaTex* and *AmsTex*, for instance). A number of my formatting hints will be given in '*LaTex* language'. However, any reader will be able to translate these suggestions into commands for his own text formatter (such as *MS-Word*).

http://en.wikipedia.org/wiki/TeX
http://en.wikipedia.org/wiki/LaTeX
http://www.ams.org/tex/amstex.html

1.F How to use the *Writing Guide*?

The *Writing Guide* is meant to be self-contained. Its study should be enough to produce well-written, well-structured papers (that still could contain bogus science).

A good, additional training in writing first-rate scientific papers is to scrutinize, with a number of colleague junior scientists a recent, short paper in a high-impact journal. These papers are supposed to represent the state-of-the-art in writing of scientific papers. This exercise might not make you happy, but it will certainly boost your self-confidence.

1.F.1 Required prior knowledge

The student is expected to have a reasonable knowledge of English, either because he is a native speaker, or because during his university education he had to study science textbooks that were written in English.

1.F.2 Further study

Some scientific organizations (examples: *AIP, APS, OSA, IEEE*) and some journals offer long and useful style manuals. (I recommend in particular the *AIP Style Manual* and an appendix of the *Review of Modern Physics Style Guide.*) The definitive information is in the Standard Handbook of the *ISO*, but unfortunately is far too expensive (you can consult an excerpt from the *ISO* guidelines). If your appetite has been wetted after reading my *Writing Guide*, you can consult any of the above-mentioned style manuals. But you should realize that you are expected to work on science and on actually writing scientific papers, and not to work too much on reading about how to write scientific papers.

http://www.aip.org/pubservs/style/4thed/AIP_Style_4thed.pdf
http://forms.aps.org/author/styleguide.pdf
http://ao.osa.org/submit/style/jrnls_style.cfm
http://standards.ieee.org/guides/style/
http://rmp.aps.org/files/rmpguapa.pdf
http://www.iso.org/iso/iso_catalogue/catalogue_tc/catalogue_detail.htm?csnumber=3653
http://www.iso.org/iso/home.htm
http://www.tug.org/TUGboat/Articles/tb18-1/tb54becc.pdf

1.F.2.A Courses in writing scientific papers

In university environments courses are frequently presented on how to write scientific papers. Do not follow these courses, even if your employer considers them obligatory. These lessons are not a good way of spending your time. The lectures are not given by active, professional researchers. Your task is to write papers that are crystal clear to your scientific community. You should cater to no other audience. The opinion of humanists, university lecturers and professors in language are absolutely irrelevant, and invariably an obstacle.

1.G Improving your English

If you want to improve your English, listen to English radio programs (like the *BBC World Service*), read high-quality intellectual magazines as the *New York Review of Books*, watch *CNN*, or watch English-spoken TV programs (without subtitles in your own language). Make sure that presentations and scientific discussions in your group are in English.

A useful and amusing resource is Paul Brians' complete website on Common Errors in English Usage.

http://www.bbc.co.uk/worldservice/
http://www.nybooks.com/
http://edition.cnn.com/
http://www.wsu.edu/~brians/errors/

1.H Culture

Much of this instruction manual is based on the assumption that the writer of the scientific texts works in a small-scale scientific group. In some branches the mores are totally different. For instance in high-energy-physics communities a four-page paper can easily carry hundreds of authors. It is obvious that writing such a paper involves, besides science, a lot of politics. And politics is (almost) nowhere to be found in the *Writing Guide*. In such communities my tutorial might still be useful for internal reports, theses and the like.

1.I Limitations

The *Writing Guide* deals with a number of different types of manuscripts. There is quite some play between the prescribed writing styles of various international journals. Some of my hints might not be allowed by your favorite journals. But these particular hints might still be useful for your internal reports, theses, proposals, etc.

1.J Male chauvinism

In many western societies women are underrepresented in the natural sciences. This absence is an undesirable situation. In this tutorial I could have been politically correct by continuously using "he/she" and "his/her". As this would make the text look uglier I have not done so. The reader should realize that wherever I say "he", it could well have been "she".

1.K Ethics

Natural science is about objective stuff 'out there'. Its findings should not depend on the observer. Work is being published to allow other workers to criticize the outcome, to check it, or to expand it.

In principle you could be requested to justify your results twenty, or more, years after the publication date. Reporting your results in detail in (internal and external) scientific texts, and archiving them extensively, apart from keeping good laboratory journals, is part of good ethical behavior of each and every natural scientist.

1.L Conventions in the *Writing Guide*

Words that are in *italic* represent (deposited) names of organizations, brands, companies and/or computer program names (and actions within a computer program). Examples: *Acrobat*, *LaTex*. Proper credit will be listed regarding these (deposited) names at the end of this guide.

1.L.1 Double quotes

Double quotation marks indicate quotes, either from text or from speech. To have the quotes stand out, their font color is red (not visible in the printed version).

1.L.2 Single quotes

In this booklet I use single quotes to indicate a 'strange' word, or a regular word occurring in an unusual meaning. In stead of single quotes I could also have used the word "so-called".

1.M Commercial products

Regularly I will mention commercial (software) products that can help you in writing your paper. If your institution is of academic nature (university, college, etc.), then for a number of these products your institute might have a (cheaper) site license.

2 MANUSCRIPT HANDLING

2.A Consistency

Before you start writing your first scientific paper, think very long about the conventions. About the names of variables. About the Greek symbols you will use for variables. About where you will put the factors of (2π) in equations. About sign conventions in your equations. Write down your conventions as a scientific text. Update your own conventions regularly and always keep them within reach.

2.B Manuscript types

There are a number of texts that can be classified as containing scientific information. Among them are:

- scientific papers,
- conference proceedings and abstracts,
- (internal) progress reports,
- theses,
- proposals for science agencies,
- manuals for your computer programs,
- referee reports,
- ...

2.B.1 Generic scientific texts

Always write any scientific text as much as possible as if you are writing a full scientific paper that can be submitted to an international journal.

When you are sending a text to a collaborator which whom you have very recently discussed your data and your graphs, this formal, elaborate writing might seem redundant and a round-about. But it is not. Laboratory journals are difficult to read by yourself and almost impossible to understand by your colleagues. Exchanging informal, incomplete scientific texts is an eternal source of confusion.

The sooner you incorporate your data and figures, complete with captions, into full scientific texts, the easier you make it for your colleagues, and for yourself in the long run (when you are writing your thesis, for instance). If finally you end up with a text that contains too many details, save a copy, make it a read-only file (with a sensible – in the English language – file name). Subsequently you can erase details and prepare the text for publication (and save it again, with another, sensible name).

2.B.1.A Extreme size constraint

A real challenge is to write papers that have extreme size constraints (three pages or so). A number of hints in this tutorial will not apply as implementation would make the text too long. Write the text first with a very good structure, and possibly, way too long. Then shorten it. In that way you are sure that no new information has to be put in right at the end.

2.B.1.B Theses

In this guide the scientific text will in the majority of cases be a scientific paper. In many aspects a thesis resembles a scientific paper. However, there are some differences.

2.B.1.B.1 Consistency

Each chapter in a thesis is like a different paper. In a number of cases this is even literally the situation.

Readers of a thesis will expect a consistent notation throughout the whole thesis. This is a serious constraint. Before you start writing your thesis, think very long about the conventions you are going to use.

2.B.2 Some special scientific texts

Below I will sum up some special scientific texts that differ considerably from generic scientific texts.

2.B.2.A Computer programs

It is beyond the scope of this guide to go into any detail on how to write scientific software. Dozens of books have been written about how to write computer programs.

It all boils down to the following: somebody else – or you yourself a few years later – using your source code should be able to use it on his own computer without much effort. Make excessive use of comments. Give variables obvious names. Do not write spaghetti-code, but use structured programming. Decouple as much as possible different functions/tasks in different files. Test them independently (if floating point, down to machine precision). Make excessive use of asserts (or the equivalent of the assert macro/function in your favorite programming language). If the assert function does not exist, program it yourself.

If you keep old versions of your computer program, give them obvious names, explain in the first comment why this version of the source code has become obsolete and make the file read-only. Show your computer program(s) regularly to your supervisor and coworkers.

http://en.wikipedia.org/wiki/Assert.h

2.B.2.A.1 Manual

If you write a manual for your computer program (highly recommended), refer in the comments of your computer program extensively to text and equations in your manual. Write the manual as a generic scientific text.

2.B.2.B Comments

Young scientists quickly get aroused agitated by recently published papers appearing on their own terrain. Papers they think are wrong. The first action that comes to their minds is to write a scientific paper in the form of an aggressive comment.

It is generally a bad idea to write a hostile comment. In the community, comment writers are viewed as complainers and querulous persons. Only consider writing a comment when you really find the work flagrantly wrong and a threat to your own scientific position. Write the comment in a very friendly, polite way. Do not use strong words. Give the people you criticize a possible way out. Otherwise you will have made enemies forever (these enemies will be the reviewers of your future proposals and papers). As a rule of thumb, a senior scientist should not publish more than one critical comment in five years.

Comments in the form of a full-size paper follow the rules of generic scientific texts. Comments in the form of a very short note are very difficult to write. Have a number of non-authors read the concept of the comment.

2.B.2.C Conference abstracts

Conference abstracts are typically very short (half a page or less). They require much less attention than regular texts. The rules are more forgiving. Abstracts are commonly set up as plain ASCII files.

2.B.2.C.1 Style of conference abstracts

Abstracts are as a rule following mild guidelines. You can do much of the (ASCII) formatting yourself. Partition a one-page abstract into several paragraphs, either by indenting or by introducing a blank line.

If you add a list of references of which some are multi-line, prevent the second and possibly following lines from wrapping under the reference numbers:

1. P.W. Jones and D. Smith, J. Chem.
Phys. 133, 345 (2010). (bad example)
2. P.W. Jones and D. Smith,
   J. Chem. Phys. 133, 345 (2010). (good example)

2.B.2.C.2 Credit in conference abstracts

It is always a good idea to give credit, in the form of literature references, to important colleagues that you expect to be present at the conference. Restricting your references to your own work only, will have a contra-productive effect. Adding literature references to your own work of the type "in preparation" is ridiculous.

The first author of an abstract must be the presenter (if the paper will be orally presented) even if he is not the actual author of the abstract.

2.B.2.D Referee reports

Referee reports are short texts (typically half a page). Always without figures, and if possible, without mathematical formulas. Many of the general hints of the *Writing Guide* do not apply.

I have included in this *Writing Guide* a number of hints dealing with how to write good referee reports.

2.B.2.E Grant proposals

The science-supporting agencies differ in their requirements with respect to formatting of the proposals.

On the worst side are those organizations (like *NWO* in the Netherlands) that require the use of prescribed style files to be used in combination with *MS-Word*. These style files are always buggy, difficult to use, and apparently always developed by non-professionals.

On the best side are those organizations (like *FOM* in the Netherlands) that describe the format and leave it to the applicant how he implements it: with *MS-Word*, *StarOffice*, *LaTex*, or whatever.

Filling out web forms is even worse. Organizations that request this as a submission channel should have their managers fired immediately.

2.B.2.E.1 Formatter

If you have some freedom in the choice of formatter for your grant proposal, *MS-Word* is probably the best. It allows for easy creation and editing of outlined text. Figures are easily imported. If you need an occasional formula (it is a grant proposal, not a math paper), you can use *MathType* or import the equation as a figure.

2.B.3 Your own layout

In the majority of cases you will be submitting your manuscript to an international journal. In such a case you have no control over the layout. In those cases where you do have freedom of layout, the simplest, professional solution is to use the style files of an appropriate journal which does have an obligatory style file.

2.B.3.A Digital format

As we are discussing those manuscripts for which you are free in the layout, you are also free in choosing the digital format. However, you have no much liberty here. You must generate a pdf version.

When you generate a pdf file you should check "optimized for online viewing" (an option in many pdf-generating applications). In addition extensive use of bookmarks should make navigation through the file a delight. Bookmarks in a pdf file can be generated from an *MS-Word* document or with the *hyperref* package in *LaTex*. A large internal report, or a thesis, without extensive hierarchical bookmarks is a mere sign of amateurism.

2.C Where to publish?

If you know in which journal you want to publish your paper, check a recent issue to find out their manuscript guidelines.

Relevant questions when you write a paper involve: how does the journal implement references? Does the journal allow partitioning in sections? Does it allow classifying of authors according to their affiliations? Does it allow singling out one author as the corresponding author? Does the journal allow a separate acknowledgement section?

If the choice of the journal is not yet fixed, the authors have to make up their minds quickly. Factors deciding the choice between journals include: number of readers, type of reader community (physicists, optical people, engineers, chemists, etc.), impact factor of the journal, time between submission and publication date, quality and speed of refereeing process, and acceptance rate.

2.D Corrections by coauthors

In the next subsection I will treat the process of incorporating the corrections suggested by one coauthor. Before this process can be started, the problem of multiple authors has to be dealt with. Suppose that there are five authors: A, B, C, D, and E. How to cope with the non-converging process of having author C undoing the corrections introduced by author B? First find out which coauthors explicitly express that they do not need to see the paper at all. These people can be immediately removed from the list of authors.

2.D.1 Bilateral correction process

First author A should make a strategic decision with respect to communication with other authors B, C, D, and E. He should not go into multilateral rounds, like BCDE in a first round, and then BCDE in a second round. Given the high ego of natural scientists, this multilateral process is guaranteed not to converge.

> Author A should pick out among the coauthors the author that is most influential, most accurate and most rapid. Say this is author C. Then in a bilateral series of rounds AC, AC, AC, a concept should emerge that both A and C fully support.

Do not indicate where the modifications have taken place in following improved versions. Any new version should be clean and free of 'historical remarks'.

This AC-concept is the basis of a series of rounds with another coauthor, say B: AB, AB. Author B should be explicitly told that A and C already fully support the concept at hand. This information puts moral pressure on author B to limit the discussions to the very necessary corrections. B should feel hindered to come up with major changes. He knows that very likely A and expert C will strongly oppose any cosmetic suggestions being brought up by B. The more authors have agreed on the manuscript the more difficult it becomes for a next coauthor to come up with unnecessary modifications.

The general idea is that it should be hard to suggest cosmetic changes and easy to make necessary modifications.

2.D.2 Checking coauthors

Author A should ensure that all his coauthors really have read the text. An immoral but very effective method to test whether or not a coauthor has fulfilled his duty is to put some deliberate, small mistakes in the concept that a conscientious reader should catch. Any coauthor should realize that not reading the concept paper carefully and not reading it within a few days could result in losing his coauthorship.

2.E Corrections by one coauthor

2.E.1 White space in concepts

A concept text of a manuscript is often sent, or handed, to a coauthor with the idea that he will make comments. For a coauthor it is highly frustrating to receive a 'dense' text. Send him a hard-copy double-spaced text with large margins. Certainly not a two-column single-spaced text.

> A manuscript with a lot of white space makes it easy to insert comments.

(To limit the physical size, you can always print two-sided, unless the text needs to be faxed back by the reviewer.) The reviewing coauthor can now very easily indicate suggestions for modifications between the lines (in red, for instance).

2.E.1.A Communication with first author

The active coauthor should not send his modifications digitally. He should just hand over, or fax, the responsible first author the paper copy containing the hand-written correction marks. A short discussion between the active coauthor and the first author where the active coauthor explains his hand-written comments, will speed up the writing process considerably. In a final stage phone calls can also be helpful to communicate with a coauthor (both should have an identical copy on their desk, or should both have the identical file on their screen). It is my experience that digitally incorporating suggestions and corrections – through comments in the pdf version for instance – is far too time consuming, and not really helpful for the first author.

2.E.1.B Size limit

Suggestions by the active reviewer to incorporate more text are unacceptable if this would result in a violation of a prescribed size limit. Unless the suggestions are accompanied by proposals regarding what parts of the original text should be left out.

2.E.1.C Final format

When the manuscript is almost complete, the format of the concept can be changed into the relevant format for the journal (for instance single-spaced, two columns).

2.E.2 Version control

There can only be one active corrector.

He has a paper copy and/or a digital version of the manuscript. Nobody else, including the first author, is allowed to introduce corrections or to change the manuscript. Tell the active corrector that all the other authors are waiting for his comments. Speed him up. But never work on a text that somebody else is also working on. Just put pressure on the active corrector or remove his active-corrector status.

2.E.2.A Long papers

The advised procedure might become unpractical for long to very long papers and books. These papers should be broken down into independent units, like chapters, and these should be treated as independent papers as far as the communication with coauthors is concerned.

2.E.2.B Meeting of authors

When you organize a meeting with (some) of your coauthors (never more than two), have all the participants bring there own hard copy. Make sure that the contents of the copies are identical.

2.E.2.C Dead line for active corrector

When you pass on your manuscript to a coauthor, agree right away on a deadline for his returning of the manuscript. A waiting time of longer than a week is unacceptable, even if the coauthor is the director of a big German institute.

2.E.2.D Incremental texts

Never supply your colleagues with incremental texts. These are texts that only contain the modifications. Reading this text fragments will require a lot of unnecessary bookkeeping for your colleagues. Just supply them with a complete, new, virginal version (with indeed all the figures as well). In addition tell them to destroy all earlier versions (paper and digital). Comments, revision history, and remarks in margins are always disturbing.

New versions should always contain a new date on the first page.

In addition, a header containing date and author/title information on each page is very handy.

It is a matter of personal feeling whether you supply your coauthors with digital or printed versions of the manuscript. If you know that the coauthor who will be the active reviewer, will print the file you send him, you can speed up the reviewing process by in addition printing it for him and putting it on his desk.

2.E.2.E Duplication problems

It regularly occurs that a scientific text has to be duplicated. An example is the situation where the same work is published twice: as a short letter article and as an expanded paper. Another example is when a detailed progress report is condensed into a regular scientific paper. Problems arise when in one of the two versions improvements are incorporated. As a result the papers will not be in synchronization any longer. This problem causes managers of software companies nightmares. Discipline is the only solution here.

Suppose that you have finished the large version, which all the coauthors agree on, and want to extract a short version from it. Make the finished large version read-only.

This will warn you if by accident you want to change this file. Work on a copy that will become the short version. Corrections to the short version, suggested by coauthors or referees, are also relevant for the large version, and should first be done in the large version. Have these improvements in the large version read and agreed on by your coauthors. Subsequently copy/paste the modifications into the short version and consult the coauthors again. It is a hassle, but diverging texts is an even greater problem.

After the paper has been published there is no duplication problem anymore. If you find mistakes or corrections in the published paper, there is no solution (in exceptional cases an erratum can be submitted).

3 TEXT FORMATTER

There are basically two types of text formatters: (i) formatters that use ASCII files containing typesetting instructions and (ii) WYSIWYG (What You See Is What You Get) word processors, like *MS-Word*, hiding the formatting in a proprietary file standard.

3.A *Tex* family

The standard text formatter for texts that contain a lot of mathematical formulas is *(La)Tex*. On the internet complete and simple-to-use *(La)Tex* distributions can be downloaded. I use *MikTex* to great satisfaction. For *Windows WinEdt* is the standard editor for *LaTex*. You can use better front-ends like *Scientific Word*, but they often include proprietary macros. You can use *Scientific Word* as clean as possible by starting it with a style file from your favorite journal, for instance an *APS-(LaTex)*-style file and saving always to standard *LaTex*. Many publishers allow submission of articles in *(La)Tex*.

> Often they encourage you to use their *LaTex* macro packages. Avoid using them as much as possible.

http://www.ctan.org/
http://www.miktex.org/Setup.aspx
http://www.winedt.com/
http://www.mackichan.com/

3.A.1 Record: length and ending of record

For *Unix* users: set record length of file to a fixed length of 72 or 80, to prevent *Windows* users from getting problems. The *Windows* operating system often inserts hard return characters when the record length is larger and these hard returns often cause *LaTex* compilation errors. (*Windows* uses \CR\LF for end of line. *Unix* uses \LF and *Apple-Macintosh* uses \CR. \CR = Carriage Return and \LF = Line Feed. They are standard ASCII control characters).

3.A.2 Compatibility

The *LaTex* 2.09 version is too old now. You have to use the newer version *LaTex2ε*. You should realize that many *LaTex* distributions have their own peculiarities (for instance with respect to the inclusion of figures). Before redistribution of your *LaTex* files try them on more than one *LaTex* formatter. Nowadays a number of websites offer on-line compile service of *LaTex* files.

http://www.ctan.org/tex-archive/info/lshort/english/lshort.pdf

3.A.2.A Converting from old versions

Turning old (2.09) *Latex* files into new ones is essentially replacing the old "\documentstyle[]{}" by the new instruction "\documentclass[]{}" and introducing some new "\usepackage{}" instructions.

3.A.3 Other text formatters like *MS-Word*

Only if you use almost no mathematical formulas *MS-Word* is acceptable. (Preprints with formulas always have an amateurish look in *MS-Word*.)

3.B *MS-Word* survival

I will outline here a minimal number of hints to survive manuscript writing with *MS-Word*.

It is quite difficult to make professional publications with *MS-Word*. This is because *MS-Word* does a very lousy job on justification. With some effort you can improve the look of a printed *MS-Word* manuscript.

http://www.aaronshep.com/publishing/WordType.html

3.B.1 Word 2007

When preparing this guide *Microsoft Office 2007* was released. Using this new version was a nasty surprise. It has been totally reworked. Much has changed. The 2007-version of *MS-Word* has become very slow, and contains quite a number of bugs and distasteful features. The GUI interface has changed dramatically.

Microsoft has introduced a number of new, proprietary file formats. You should not use them for the time being as the majority of your colleagues will not be able to open and read files in these new file formats.

Until further notice I advice to adhere to an older version, like *Office 2003*.

If you are pretty good with editing the *Windows* registry database, then you can keep a new and an old version of *MS-Word* simultaneously on your computer.

3.B.2 Justification

The only acceptable justification (alignment) in *MS-Word* is left aligned. You will have a ragged right edge. This is not unprofessional. It is just not as appealing as full professional (full) justification.

3.B.3 Left alignment

If you apply (*MS-Word* default) left alignment, you can reduce the raggedness considerably with thorough use of hyphenation (see next item). In the hyphenation process words get invisible soft hyphens, possible points where that particular word may be hyphenated. (Unfortunately an ever surviving bug in all versions of *MS-Word* disables the power of the spell-checker if words contain soft hyphens.)

3.B.4 Full justification

If you want to advertise your dilettantism and insist on applying full justification (justified alignment: flush left and right sides) you can limit the damage if you apply the following hyphenation procedure.

To prevent the occurrence of too much white space (or to prevent the ragged right side from looking too ragged in case of left alignment), do the following when a new version of your concept is almost ready:

> make the left and right margins temporarily very large – the text width will be very small now – and start hyphenation (*Tools, Language, Hyphenation, Manual*); continuously accept the choice *MS-Word* offers you.

After possibly hundreds of quick mouse clicks (or keyboard actions) many words will have obtained soft hyphens. Restore the original margins. Soft hyphens can, if needed, easily be removed with *Replace*.

If you apply full justification without hyphenation, you deserve to be banned.

3.B.5 Mathematical equations

MathType is a plug-in for *MS-Office* that considerably improves equation editing (much better than the *MS-Office* Equation Editor).

If the journal you are submitting your manuscript to tolerates equations being introduced as figures, then you are lucky. You can prepare your figures with a professional formula formatter, like *Tex*, *AmsTex*, or *Latex*, and export each equation as a figure.

http://www.dessci.com/en/products/mathtype/

3.B.6 Recovery

> Save the file you are working on continuously (click the floppy icon, or type Ctrl-S). Do it at least once every five minutes.

AutoRecovery often does not work properly (it has been improved in *Office 2003*). If *MS-Word* freezes and you know from experience *AutoRecovery* will fail, then do the following: exit all other programs and switch off the computer power. Following this procedure ensures that the temporary autorecovery files are not deleted. Restart your computer and restart *MS-Word*.

3.B.7 Navigation

Use the extensive outline facilities of *MS-Word* (*Document Map*). If you switch on *Document Map*, the whole structure (heading, subheadings) must become visible. In the beginning it requires some trial and error to find out how the outlining facilities of *MS-Word* work, but later on you will be very happy.

Customized automatic numbering of the outline is possible.

Users of your text can now easily navigate. Tell them of this navigation possibility because the world is full of ignorant *MS-Word* users. Maintaining and improving the text also becomes much simpler.

When exporting to pdf you might want your numbering of headings to be exported to the pdf file. Here you can find how you can do this:

http://pubs.logicalexpressions.com/Pub0009/LPMArticle.asp?ID=551

3.B.8 Dictionaries

The included Thesaurus (certainly in the versions up to *Office 2000*) is too restricted. Use your own. See the section on dictionaries further on.

3.B.9 Macros

Only use very simple macros. When you put dates as a macro in your manuscript, lock them (CTRL+F11), so that reopening the file does not automatically update the date.

3.B.10 Templates

Avoid complicated dedicated templates (dot files).

If your granting organization prescribes them and supplies you with their own template (always amateurish, usually with numerous horrible tabs for alignment, introduced by ignorant managers having followed a half-day course in the use of *MS-Word*). Try to edit their template and remove all frills. Try to erase the macros in their template and implement the formatting your own way.

3.B.11 Version control

Do not use the possibility to save multiple versions of the manuscript. It will only give rise to confusion between coauthors.

Refrain from using the ridiculous track changes with color and strike through in *MS-Word*. Use a conspicuous set of characters, for instance "@@@", or "****" to start and end your comment or suggestion for change. They can be found easily, printed easily, and removed very quickly without laborious 'unformatting' mouse and key actions.

3.B.12 Wid ows

Preview the print version before printing and introduce extra hard page breaks to exclude the occurrence of ugly 'widows and orphans' (headings that start on the bottom of a page). The built-in facilities to avoid these widows are often not satisfactory.

4 TEXT STRUCTURE

4.A Organization of content

4.A.1 Margins

Regularly journals prescribe a margin size.

If you are free to choose your margins, choose them pretty large (left, right, top, and bottom).

This gives a relaxed impression and collaborators can easily insert comments in the printed version of the text.

4.A.2 Justification

Left and right margins determine where a line starts and where it ends. In texts four forms of justification can be applied:

Left justification. Lines have about equal length (if not overruled by hard returns or manual line breaks). New lines are aligned with the previous one on the left side. Hyphenation can reduce the raggedness on the right side. Also referred to as left alignment.

Right justification. Lines have about equal length (if not overruled by hard returns or manual line breaks). New lines are aligned with the previous one on the right side. Hyphenation can reduce the raggedness on the left side. Also referred to as right alignment.

Centered justification. Lines have about equal length (if not overruled by hard returns or manual line breaks). The center of new lines is aligned with the center of the text width (space between the margins). Hyphenation can reduce the length variations of the various sentences. Also referred to as centered alignment.

Full justification. There are no leading and no trailing spaces. Unprofessional word processors, like *MS Word*, implement this type of justification by adding extra space between the words in a sentence. Professional desktop software, like *Adobe's InDesign*, achieve this by adjusting the space between individual characters, by adding (micro)spaces between words, and by micro-changing of font sizes. Hyphenation can reduce the variation in character density of the various sentences.

The *La*(*Tex*) family uses a reasonably professional form of full justification.

You have no control over the justification used in a professional scientific journal. You do have control if you are writing an internal report or a thesis. If you do have control, never use full justification the *MS-Word* way. Either do it professionally or use left justification.

4.A.3 Subheadings

You write to be read. To be cited. To be followed. To be criticized. Help your readers. Break your results down into bite-size pieces. Keep it simple.

Use as many subheadings as possible.

The reader then knows much quicker whether or not he can skip the section. You can even explicitly help the reader to pass over some of your sections: "Readers not interested in the mathematical details can directly go to Section X".

Number all your subheadings if the journal allows it. This numbering will facilitate communication between multiple readers of your paper.

Try to minimize the coupling between different subsections. The lesser a subsection depends on the content of other subsections, the quicker your paper can be (partially) read.

If you find the number of subheadings too large, you can make them 'temporary' and remove them later, bij 'commenting' them out.

4.A.3.A Hidden subheadings

A class of scientific papers do not allow to have subheadings in the paper. A number of short-letter type papers belong to this category.

Inexperienced authors should nevertheless use subheadings in these short notes. In the last stage of the preparation process these subheadings can be removed, or better: can be 'commented out'.

4.A.4 Paragraphs

A paragraph is a natural unit of information.

A paragraph containing more than ten sentences is often too long.

Look for a natural breaking point in a large paragraph. There you can start a new paragraph. Paragraphs are anchors for experienced readers. Help those readers to be able to quickly scan your paper.

4.A.5 Spaghetti text

Do not write spaghetti text by continuously referring to earlier pages or formulas. Readers get tired when they have to leaf through a text often. Your public is not supposed to consist of pathfinders. You are neither writing for the referee, nor for somebody who is going to spell your text. Your average reader is mildly interested, and continuously looking for an excuse to put your paper aside. The use of headers and footers considerably increases the spaghetti-ness of your text. Avoid these notes as much as possible. Exceptions are of course footers and headers that contain page numbering or title/author information.

4.A.6 Parking places

Some subsections are parking places for information. For instance the subsection, or the part of the paper where the experimental details are presented, should be complete. Nowhere else in the paper should the reader be bothered with these experimental details. So a figure caption containing "The detector was calibrated using ..." is a crime. By putting all experimental details in the experimental section, the rest of the paper will be much easier to read and the interested reader does not have to use a radar device to find the spot where he can read about a particular experimental condition.

4.A.7 Standard partitioning

Many of your readers, including Nobel Prize winners, like and expect the standard pattern: introduction, experimental details, results, discussion, conclusion. Do not try to be original here.

> Do not surprise the reader with an original structure.

4.A.8 Smooth text transitions

A scientific text should be a logical sequence of arguments. Nevertheless transitions will be present. Make them smooth. Announce next sections if they constitute a transition "In the next section we will present our new model".

4.B Length of sentences

Try to keep sentences short. A sentence longer than thirty words is difficult to read. You are probably a non-native English speaker and your reader certainly is. It requires quite some skill to produce easily readable sentences containing more than one comma. Always try to replace a comma by a period. Always try to replace a dependent clause that starts with "which" or "that" etc., by a new sentence.

5 TEXT CONTENT

The first author is in some sense the moral owner of the text. Coauthors should correct and come up with suggestions to improve the text. If in the end there is a discussion about language variants that differ only in taste, the opinion of the first author prevails. Senior authors should restrain themselves and not impose their own taste. First authors should be persistent in not giving in.

5.A Obligatory items

5.A.1 Title

Start with a working title.

Finalize the title only at the end.

Titles are short. No definite articles. Be precise. If your paper is 100% experimental, do not use a title that might give the impression that you have also done theory.

5.A.2 Authors

5.A.2.A First author

The first author writes the manuscript. Other authors comment and give suggestions. Never should a senior scientist write part of the paper himself. He should keep on discussing and suggesting modifications until he can live with the manuscript. A senior scientist, not being the first author, who writes part of the manuscript himself, will seriously depress and indeed insult the first author.

5.A.2.A.1 Many first authors

Group leaders with a big ego organize the writing of a paper differently. They split up the content of the paper in a number of parts. Each prospective coauthor gets to write such a part. They all have to turn in their sections and the group leader will glue it all together. Junior scientists in such a group will have a hard time growing up.

5.A.2.B Coauthors

Who will be coauthor? If your mother did not teach you any decency it is too late now anyway. The question of coauthorship only arises when the text is mature enough to be a concept for a paper. Or when there is agreement on what material should be in the paper. Discussing authorship earlier might give unnecessary friction, because the question who is going to be an author depends heavily on what is finally going to be in the manuscript.

Only put names of authors above the concept if you are sure that the list is complete.

Sending a concept with authors A, B, and C to colleague D with the question whether or not he wants to be a coauthor is an insult to colleague D.

5.A.2.B.1 Placeholder

If you want to postpone the discussion about the list of authors, circulate the concept with as authors the name of the first author and the placeholder "plus additional authors". Prospective authors that got a chance to glimpse at the concept do not feel excluded (yet).

5.A.2.B.2 Individual contributions

Some journals request a statement regarding the individual contributions of authors. The only acceptable solution is: all authors contribute equally. The reason being that it is very difficult to compare contributions. How would you weigh the 10% time input of a highly experienced, creative senior scientist against the 90% time input of an inexperienced junior.

5.A.2.B.3 Contribution of world specialist

Each and every author of a paper should understand the basics of that paper. Each and every author should be able to give a presentation on the content of that paper. With an increasing number of multidisciplinary papers this moral standard becomes more and more difficult to uphold.

A real moral problem is the specialist who has something that is indispensable for your research – a sample for instance. The specialist understands nothing of your work, but he requests to be coauthor on all the papers where you publish results obtained with his sample.

A good compromise is to put him on the first paper – sandwiched between a number of other coauthors – and leave him out on all future papers.

5.A.2.C Order of authors

The natural order of authors is: first author is the person who has spent most of his time on the subject, often a PhD student. The last author is the group leader. The remaining order is alphabetical between the first and last author. Exceptions can be if there are more main contributors. Realize that readers will never understand subtle changes you make in the sequence of the list. In some journals you have to group the authors according to affiliation. In such a case the affiliation of the main contributor will be the first listed affiliation.

5.A.2.D Corresponding author

The role of the corresponding author is not clear. Many group leaders want to be the corresponding author. If they would leave this task to the (junior) first author the seniors feel they will 'lose control'. I think this is a bad habit of group leaders.

> The first author should be the corresponding author, unless he has already left the group.

If questions are posed about the content of the paper, the first author is most likely to be able to answer them best. The corresponding author should never answer the question without consulting the coauthors. Correspondence should always be cc'd (carbon copy) to all authors. This informing of coauthors even holds when the group leader is the corresponding author.

5.A.2.E Spelling of names

Beginning scientists should make up their mind about how they want their names be printed all through their career. I am referring to spelling, the number of initials, use of first name, leaving out or maintaining difficult characters (like "ü", and "ó") and more.

Whatever you choose, be consistent all your life. If you fail to be consistent, citation databases will classify your papers under different names. You will get a much reduced citation record. In addition people might get confused about whether or not "Bill Smith" is the same as "J.W. Smith".

Europeans should realize that there can be funny misconceptions, especially with the use of capitals in name additions. A name spelled as "Van Beethoven" is to an American a guy whose first name is "Van".

5.A.2.E.1 Use of first names in author list

Each author in a list of authors can decide individually whether or not he wants to include his first name. Some journals might refuse to print first names.

5.A.3 Affiliations

The first concept of the paper is always written by the first author. This writer will in many cases be a junior scientist without much experience. For him the correctness and completeness of author affiliations will not be important.

Since the 1980's names of institutes change regularly, often following fashion and hypes. For instance since the year 2000 more and more institutes have incorporated the prefix 'nano' in their new or renewed names. To keep the peace with local managers and operators, it is important for seniors to check that the list of affiliations is correct, actual, and complete.

Seniors should be careful if they have dual appointments.

They should give full credit to all those employers who have supported the reported research.

5.A.3.A Change of affiliation

The affiliation reported in the paper should be the place where the work has been performed. Junior scientists change their affiliation after they have finished their PhD. They might write papers about their (old) PhD work while already employed at a new place. Such junior scientists like to please their new group leaders by trying to get their new affiliation on the paper. This is immoral. The old institute has invested a lot of money in the research without getting any credit, whereas the new institute would get a paper for free.

5.A.4 Abstract

First write the paper and then write the abstract.

Be very precise about the character (experimental, theoretical, numerical simulation, synthesis, etc.) of the work you have accomplished. Do not mislead the reader. Help him.

Indexing in a number of (internet) databases will be done on the basis of keywords found in your abstract.

No math, tables or references in the abstract.

The abstract will usually be followed by either a list of keywords or a few keys indicating subject classification.

5.A.4.A Keywords

Be as precise as possible. Do not try to attract readers on false pretexts. Copy the keywords out of a paper of your successful competitors, or copy them out of an older paper of your own group.

5.A.4.B Subject classification

Be as precise as possible. Do not try to attract readers on false pretexts. Well-organized societies, like the *APS*, maintain a comprehensive list of subject classification that has to be used on all papers submitted to these societies. In the case of the *APS* it is called the PACS (*Physics and Astronomy Classification Scheme*).

Copy the subject classifications out of a recent paper of your successful competitors.

http://publish.aps.org/PACS/

5.A.5 Date

Take care that the output, hard copy or digital, will contain the date (at least once on the first page 1).

Of course not with "\today" (because this *LaTex* macro will be updated each time you recompile the *LaTex* file). Very convenient for your coauthors is the use of "\pagestyle{myheadings}" left and right with the date. Every page will contain the date now. If one of your coauthors has two different hard-copy versions that got mixed up (occurs very regularly to me) they can immediately separate them again, and throw away the oldest. When you submit the file to a journal, this heading feature can always be commented out. The date should be the date of the last revision, or of the submission date. Following is an example (the use of \hspace is for aligning).

```
\pagestyle{myheadings}
\markboth {\underline{\hspace*{1cm}Megens et al.
 { 'Fluorescence lifetimes and  linewidths ...'}
  Submitted to Phys. Rev. A. on 11 Nov 2007}}
 {\underline{\hspace*{1cm}Megens et al.
 { 'Fluorescence lifetimes and linewidths ...'}
 Submitted to Phys. Rev. A. on 11 Nov 2007}}
```

5.A.6 Introduction

The first line of the introduction, which is the first sentence of your paper, is a very important part of your paper. The first paragraph will to great extent determine whether a potential reader will stay away from the rest of your paper.

Get inspiration by reading the introduction of some recent papers by a successful competitor. You could even copy it and change it by using some synonyms.

Use the introduction to describe the status of the field.

A little 'hyping-up' is desirable ("great interest", or "dramatic developments", "has become a very active field of research"). Do not present your results in the introduction. Some history might be appropriate.

"This paper is organized as follows" is a cliché.

5.A.7 Conclusion

A conclusion is not a summary. In a conclusion you sum up your findings. Not what you have done.

For a short paper this might not be necessary. Avoid repetition of text already present in the rest of your paper (like the introduction).

A common mistake is to present new information in the conclusion.

5.A.8 Acknowledgement

You can be sure that your local colleagues will spell your acknowledgement section. Never forget the acknowledgement section. Do not circulate concepts of your paper without a full acknowledgement section, or put a place holder there "we thank (to be completed) ..."

5.A.8.A People in acknowledgement

Forget nobody. You can be sure that all your group member will scrutinize this section. Just be gentle to people that helped you. Next time they will help you again.

Present names in alphabetical order, unless some people made special contributions. Be consistent with first names: all names with first names or all names with initials. Do not use titles. If your boss is a German professor convince him that in American English he does not lose his authority if you use his first name. Under all circumstances you should avoid German, Dutch, and Belgian titles like Prof. dr., Prof. dr. habil., and Prof. Dr.-Ing.

In American English usage of a first name does not necessarily imply informal relations.

5.A.8.B Refused coauthors

Some people might feel very uneasy about the fact that they are not a coauthor. You can de-escalate this situation by spending a warm paragraph on their role in the acknowledgement.

5.A.8.C Institutions in acknowledgement

The first concept of the paper is always written by the first author. This writer will in many cases be a junior scientist without much experience. For him the acknowledgement section will not be important.

A scientific group gets (some) support from its host institution. This support is acknowledged through the mentioning of the affiliation in the title. No extra acknowledgement, please.

Much research is in addition supported by science supporting agencies. These life-saving organizations would like to be recognized, often in a prescribed way. To keep the peace with those organizations, the acknowledgement should contain this information exactly according to their request. For juniors: just look at some previous papers of your group to see how the acknowledgement section should be written.

5.A.9 List of references

Together with the acknowledgement section, the list of references will be the best studied part of your paper. It is there where you make friends and enemies.

The enumeration of papers will to a large amount determine how referees and colleagues receive your paper.

5.A.9.A Choice of references

Be honest with your choice of references. Papers or books that were essential for your own research should be cited. Give them ample credit and refer to them more than once. Even if you hate the authors.

> If your field has a number of leaders, try to include in the list of references at least one paper of each leader.

If a reader will start to browse through your paper, the first section he will probably consult is the list of references. Very likely, to check whether or not you referred to his work.

This rushing to the list of references is also an act that the referee will perform. If the referee feels you have neglected his work, both you and the referee are in an awkward position. You, because you will be confronted with a hostile referee, and the referee because he cannot bring up this point in detail as he will risk blowing his anonymity.

Size limits on the paper might seriously constrain the number of references you can include. In case you have to make a choice between skipping a real reference or a reference to ease a potential referee, omit the real reference.

5.A.9.B Blockbuster

If your field has some extremely well-cited papers, cite them as well. It will make your field more influential.

5.A.9.C Reviewers

When the editor receives your paper, he has to choose referees. He will very likely pick at least one, if not all, of the referees out of your list of references. Be aware of this important aspect of the list of references.

5.A.9.D Own work

Citing your own work is only allowed if you do it in a modest way.

> Not more than 15% can be self-citations.

A higher percentage will be interpreted as a real sign of dealing with an author that considers himself a misunderstood genius.

5.A.9.E Unpublished

Referencing unpublished (like "unpublished" or "private communication") material is a bad habit. It shows something that a real scientist is not supposed to do: he is bragging about information he has and nobody else has. Moreover this secret information is apparently important for understanding the paper.

If the reference to unpublished work is to your own work or one of your coauthors, you open a dangerous can of worms. It may give the referee the idea that you are still working on the subject and that it is not finished yet. Especially when you are trying to get your paper published in a high-impact letter journal, you make life unnecessary hard on yourself. The referee will say "Why don't the authors write an expanded paper on the subject where all the unpublished material can be included". This type of argument by the referee is a real killer.

5.A.9.E.1 Internet

Often the unpublished material is not really unpublished. For instance regularly theses are cited as unpublished. But nowadays there is no excuse whatsoever for not having these theses online on the internet. In such a case the uri of the thesis can be used as a real reference. This is beneficial in many ways. The thesis might be read much more than when not online and you give every reader (and referee) in principle the possibility to check and browse, or maybe even read the thesis.

5.A.9.F Preprint server

If you want to refer to work that has not been published yet, there is a solution: submit your paper to an established scientific archive.

For instance the *Cornell University* archive, arXiv, which is an e-print service in the fields of physics, mathematics, non-linear science, computer science, and quantitative biology. These archive papers are not refereed, but as they are available to everyone, they can be cited.

http://arxiv.org/

5.A.9.G Accepted for publication

It is legal to cite your own work that is in the process of being published. Mention the journal where it is going to appear ("Accepted for publication in Chem. Phys. Lett."). Be careful that the referee can use this as an excuse not to accept your present paper. The referee does not know the content of your paper and he can guess that it has a lot to do with the present paper.

Apparently you are, according to him, in the game of 'serial publications'. Or as the *APS* puts it: After a paper on whifnium we do not need a paper on whafnium and later one on whoofnium. See the very interesting *APS*-text on referee criteria.

http://prola.aps.org/pdf/PRL/v28/i6/p331_1

5.A.9.H Private communication

Private communication as a reference is only acceptable if you use the results of a colleague that he is going to publish later.

In this sense it is equivalent to "to be published". It must be information that an interested reader should be able to get hold of eventually.

If the private communication is essential for your paper and will not be published, your style of writing is immoral.

5.A.9.I Non-journal references

A short comment that would, if put inline, hamper the flow of reading can be put in a reference. Like: Reference 10: "Our convention for the Hilbert transform differs by a factor of 2π from the convention in Reference 6."

5.A.9.J Foreign language

Citing papers that are not in English is futile.

The majority of readers can not study these papers. These foreign-language citations often serve as a claim of priority. As a referee I never accept these foreign-language references in a paper that I review. If there is an English translation of the journal, it is OK, but then the English translation should also be given in the list of references.

5.A.9.K Spelling of names

Check, and check again, that all the names are spelled correctly. Referee A. Kühn will be utterly displeased if you spell his name as "A. Kuhn". Never, ever use first names in the list of references.

Check the citation database *ISI Web of Knowledge*, or *Elsevier's ScienceDirect*, or just *Google* the name with some extra keywords, if you are not sure about the spelling of the names. Do not be sloppy. Authors will feel insulted if you misspell their names. Referees will feel insulted. Why make all these enemies, when by paying attention and being accurate you can make friends?

http://www.isiwebofknowledge.com/
http://www.sciencedirect.com/

5.A.9.L Use of "et al."

To shorten the list of references, a long list of authors can be shortened by et al., like "First Author et al. have discovered …" The abbreviation "et al." is short for "et alii", a Latin expression for "and others".

Only use et al. if otherwise the author list will be much too large. Do it consistently: if the number of authors is larger than a certain number (for instance three) use "et al." A real bad example is to use "et al." for your competitors and use the full list for self citations.

Journals might have a policy about the maximum number of authors above which "et al." has to be used.

5.A.9.M Use of "ibid."

If your reference entry has more than one paper by the same authors and if the journal name is long, then you could, instead of repeating the journal name, use the abbreviation "ibid.".

This is Latin for "ibidem", which means "in the same place". It certainly will impress referees (if they are well educated).

If the same author has published more than one paper and you refer to them, do not repeat his name, just use a semi-colon to separate the papers.

5.A.10 List of figure captions

Some journals request that the submitting author supplies a list of all figure captions separated from the figures. Later I will describe in detail how to write a figure caption. Separate very clearly the various captions (by several line spacings).

Do not let the caption text wrap around the figure number. The text lines should be left aligned at a position right of the figure number.

5.B Experiment

If your paper is about reporting the results of experiments, you should state as many facts as possible about your work. Readers should in principle be able to repeat your experiment or find possible weak points in your set-up.

Never hide important aspects of your work in order to prevent colleagues/competitors from catching up.

Experimental data should always have error bars. If you present error bars as a percentage, "5 cm ± 5%", the relative magnitude of several error bars, even if they relate to totally different observables, can be compared quickly.

5.B.1 Arbitrary units

When you have to use arbitrary units, make them as 'unarbitrary' as possible. For instance, two figures both having arbitrary units, might be compared with each other in an absolute way. So relate different arbitrary units in one paper as much as possible. Arbitrary units are regularly abused in scientific papers.

The abuse of arbitrary units in experimental papers is getting a plague. Experimentalists use them abundantly to hide aspects of data they do want the reader to know about. They use them to force the conclusion on the reader that the new data seem to agree very well with theory and other experiments.

Equally bad is the presentation of scaled dimensionless absolute numbers. Suppose a dimensionless observable can vary between 0% and 100%. The maximum 100% is the theoretical limit and suppose that the actual experimentally obtained number ($\leq 100\%$) is a sign for the quality of the experiment. If the actual maximally observed number is 80%, I have seen in a number of cases the Y-axis being rescaled by immoral scientists, so that 80% became 100%.

5.C Theory

Readers should ideally be capable of repeating your theory and checking all your equations.

5.C.1 Arbitrary units

Never use arbitrary units when presenting the results of theory. By using arbitrary units you make it impossible for a new worker to check his theory against yours. The use of arbitrary units comes from the fact that experimentalists often can only determine the relative magnitude of a property and not the absolute magnitude. A theoretician should not scale his theoretical results in order to get agreement with theory in a figure.

Instead the axes should contain the absolute units of the theory and the experiments should be rescaled such that the agreement between theory and experiment is maximized.

5.C.2 Proportionality sign

Never use proportionality characters (like "$\propto$").

Usage of them means either laziness on your side or secretiveness. Rather than using "$y \propto t$" you should write "$y = ct$", and give at least the dimension of "c". Explain why you (and the reader) cannot know its value, for instance because it depends on a number of irrelevant, not-known experimental details.

5.C.3 Simulations

The use of the computer for simulations of physics systems has become very important. In principle these methods classify as theory, but in practice – due to the use of random numbers – other workers can only reproduce the results within the statistical errors of the simulation. This uncertainty gives simulations an experimental flavor.

Workers in this field are notorious for holding back essential information from the reader. Basic input as the number of particles, the number of configurations, and the size of the error bars should be reported.

5.D Theory and experiment

If your paper is a combination of theory and experiment, you must take care to separate the two ingredients as much as possible. If your competitors later discover that your experiment was ill-performed they will rightfully publish that information. But that criticism does not mean that your theory is wrong.

> So report the theory as general as possible and apply it to your experimental outcome in a separate, later section.

Progressing in this way will ensure that your theory will survive, even if your experimental part will not.

People might find one or more mistakes in your theory or in your interpretation of your experimental data. If you did not disconnect the experimental part well enough, the community will consider your whole paper to be flawed.

5.D.1 Your new model

Your theoretical paper might contain a new model, but also a description of the salient features of an old model. To have your new model stand out, you might have to present aspects of the old model. Now you have a problem: if you describe the old model in some detail, your paper becomes much more didactic. But you reproduce already-known stuff. If you give only reference to the old model and no description, your paper will be much harder to read.

If size constraints allow, present the features of the old model as well. But make very clear that you have contributed nothing to the development of the old model. Separate your model very much from the old model (use for instance "their" and "our"). Give extensive referencing to the old model.

5.D.2 Message to theoreticians

> If your paper publishes experimental data, make your work attractive for theoreticians.

Supply clear, simple tables and/or figures, so that theoreticians know what they have to calculate. Do not expect theoreticians to apply all kinds of complicated transformations and conversions to your experimental data before they can compare your data to their own theory. Try to speak the language of the theoreticians.

5.D.3 Message to experimentalist

A real killer for a theoretical paper is the following ending sentence: "We believe that the formalism presented in this paper, will eventually be useful for the interpretation of experiments". Do not convey complicated mathematical equations alone.

> Illustrate your theory with realistic examples, shaped in tables and figures.

Do not use unrealistic values for parameters: magnetic fields of 1 million Tesla, temperatures of 10^{-15} K, and an optical index of refraction of 500. If possible use values that are readily experimentally realizable.

If you present your theory well, experimentalists will be stimulated to work on your suggestions.

5.D.3.A Scaled units

Theoreticians like to reduce their parameters to a small numbers of dimensionless numbers. There are a number of good reasons to do it.

However, scaled units should always be converted in a theory paper to real numbers for realistic examples. In such a way quite a number of theory papers would never have been published, as the used values for the dimensionless units refer to utterly unrealistic values for real experimental parameters.

5.E Priority

> Make clear what are the new things in your paper.

When discussing your own work, constantly refer to "our (my) results" or "our (my) findings". When treating the results of other groups, refer to "their [reference #]", etc. Never make the impression you claim the results of somebody else. Make this separation very carefully.

5.E.1 Credit to others

You will regularly give reference to previous work by others and by yourself. The highest credit you can give is by name: "Johnson [4] was the first to point out the importance of the temperature. ..." A less prestigious form is "The importance of temperature was recently pointed out.[4]" Note the position of the reference key after the period (not allowed by a number of journals), otherwise put it directly after the reference name: "Johnson [4] discovered the effect and Smith [5] explained it."

Numbering should be according to the first appearance in the text. Sequences like [4,5,6,7] should have been replaced by your software by [4–7]. If they have not, do it by hand.

Be honest with giving credit. Furthermore this will help to get your paper accepted. Positive comments accompanying references to previous work will help you make friends forever. Examples: "In a pioneering paper John Doe derived the equation we will use here extensively [Reference #]", or "This important conjecture was put forward for the first time by John Smith [...] [Reference #]", or "In an important paper [Reference #] Jones [...]." With this courtesy you will make friends. And you will need friends, among others, because a referee had better be a friend.

5.E.1.A Criticizing prior work

Your results, experimental or theoretical, might seem to contradict previously published work. Be extremely careful how you deal with this. Expect all published work to be honest accounts of professional colleagues.

Sentences like "In contrast to Reference [2] we did not find ...", are dangerous and ambiguous. It will always be read as a debunking of reference [2]. Whereas you might want to say that the results are different, but could be explained by performing the experiment under different conditions. If there seems to be a possible discrepancy that could be due to different conditions, do not state the discrepancy without mentioning the possible cause.

> If you are really sure that under exactly the same conditions you could not reproduce previously published results, you should state "We could not reproduce the results regarding [...] as reported in Reference [2]".

The remarks in this paragraph still hold if your criticism applies to previous work of your own group.

5.E.2 First-time claim

Some high-impact journals expect sentences appearing that contain words like "For the first time we ..." Use them scarcely, overuse of "new", "novel", "for the first time" is irritating for the reader and referee.

5.E.3 Emphasize own work

A major reason for accepting your paper could be that it implies a considerable improvement over earlier published work. When comparing your own results with prior work of other groups it might help if you downplay their work and do some 'playing up' of your work. It is all a matter of language, "Elliot et al. could not get to lower temperatures than 1.2 K", whereas if you would describe your own work: "We could reach pressures as high as 340 GPa".

Be proud of your work. So often use possessive pronouns like in "our model", "my derivation", "our findings" instead of "the model", etc.

5.F Grammar

5.F.1 Use of "This means" and "That implies"

Avoid using vague referrals like "This means", "That accounts for", or "From this we conclude".

These words, without further specification are without exception ambiguous. There is always more than one sentence, statement, noun, and/or verb in previous sentences that could be the target. So always avoid using standalone words like "this", "that", "there", "where", "latter", or "former". Add at least a noun: "this observation means", or "that break-down implies".

5.F.2 When "that" or "which" leads a clause

"That" is for use in a restrictive clause: "The particles that have negative energy ..." Apparently there are particles that have negative and particles that have positive energy, and we want to restrict ourselves to the ones with negative energy. "Which" should be used if the clause is not essential: "The particles, which have negative energy, ..." There are no particles with positive energy. If you can put a comma before "that" it should have been "which".

5.F.3 Same words over and over again

Do not continuously use "results" and "resulting in results" and so on.

Or "show", "to be shown". Or "systems". Do not apply trites like "Now" and "Thus". Use "However" sporadically. It is tempting to use terms like "Now" because your whole article is one line of argumentation. Use a Thesaurus to find synonyms.

Do not write literature. Do not pretend to be Nabokov. Many of your readers are non-native English speakers.

5.F.3.A Thesaurus

Use a Thesaurus on cd-rom. Buy one that you can fully run from your computer (that is without the necessity of having a cd or DVD in the CD-DVD-reader of your computer). Or buy in addition a computer program that emulates CD-ROM drives. (for instance from CD-ROM Emulator)

http://www.cdrom-emulator.com/

5.F.3.A.1 Online

There are a number of online Thesaurus websites these days (examples Thesaurus.com and Merriam-Webster). Their working can be slow and their interface is full of ads.

http://thesaurus.reference.com/
http://www.merriam-webster.com/

5.F.4 Absolute statements

Absolute statements are often (95%) ambiguous. "When the wavelength is large" is vague. For one physicist a Planck length (10^{-35} m) is long, for another a kilometer is long, and for yet another a light year (9,460,730,472,580,800 m) is long.

Try to be more specific: "When the wave length is much larger than the size of the object". Or "Distances far away from the object" is ambiguous. Use "Distances much larger than the wavelength" or whatever length is appropriate.

5.F.5 Exclamation marks

Never use them! A scientists that studies your paper, gets the feeling that you are yelling at him. (In addition there can be an ambiguity with the factorial sign, the use of which you should also try to avoid by the way.)

5.F.6 Emphasized and underlined text

There are several ways of emphasizing text. One way is to put the words in *italic,* or **bold**, or ***both***. Another way is to underline part of a sentence. Employ emphasized text seldomly. The use has a highly patronizing effect ('schoolmasterly'). Underlined text is just plain ugly. Bold words in a regular text are also repulsive. Utilizing italic fonts for different purposes, like indicating a brand name, is allowed.

5.F.7 Active or passive form

Too much application of active form can look arrogant. If you are the only author, too much use of "I" will also appear arrogant. There can be a difference in meaning between active and passive form: "We performed the following experiments" rather than "The following experiments were performed", but also: "The crystals were purchased" rather than "We purchased the crystals" (in the latter case the authors really went to the shop). "We" is allowed for a one-author paper if it means: the readers and I.

When using passive form, avoid using the present tense: "The experiments are performed ..." The best is "The experiments have been performed". The last sentence is much more 'definitive'.

5.F.7.A Abstract subjects

Avoid sentences like: "This figure proves ..." Use instead "From this figure we deduce ..." You are the actor and not the abstract subjects.

6 TEXT SPELLING

6.A Consistency

Use consistent spelling throughout. Realize that English spelling is different from American. The majority of English journals will use American spelling ("center" and not "centre", "defense" and not "defence", "polarize" and not "polarise"). Distributing concepts of texts with spelling errors means that you wish to abuse your coauthors as spell checkers. Or even worse: you really wish to make an editor and/or reviewer angry.

6.B Hyphens

Be consistent with hyphens. Many non-English languages (like Dutch and German) differ considerably from American-English with respect to the use of hyphens. A good (American) rule of thumb is: never concatenate two nouns, never use a hyphen (or dash if you wish) unless it is an adjective. Example: "the Green function" (or "the Green's" function) and "the Green-function method". Realize that even among native English speakers the use of hyphens is a matter of debate.

6.C Quotation marks

Always use only one sort of quotation marks for quotations.

6.D Full capital abbreviations

Get around as much as possible to defining full capital abbreviations (like RPA, SPP, FWHM, etc.). They distract and make the text less transparent.

Repeat their meaning regularly in long texts.

6.E Conventional abbreviations

A number of terms will occur over and over again. In physics texts examples are: "Figure" and "Equation". In many journals they are abbreviated to "Fig.", "fig.", "Eq." or "eq.". Check with a recent issue what the style of your target journal is. It will be received as sloppy by the editor – and possibly the referee – if the standard abbreviations you use, are incompatible with the practice of the journal.

6.F Non-alphabetic characters

Usage of non-alphabetic characters like "&" and "@" in text is an indication of a high nerd value of the author. Just use "and" instead of "&" and "at" instead of "@".

7 MATH

The amount of math in a science paper can vary from almost absent to a paper consisting solely out of theorems and equations.

7.A Conventions

You probably will write many scientific papers in your life.

Try to be consistent with math symbols in all your papers.

So not "n" for density in paper 1 and "ρ" for the same density in paper 2.

Think at the beginning of your career about a number of conventions, like the sign in Fourier transforms, the location of "2π" in a Fourier transform.

Make a choice and stick to that choice as much as possible. A number of your papers will probably be the basis for your thesis. Different conventions in the various papers will make it much more difficult to combine the papers into a thesis.

7.A.1 Alphabet

When representing a property by a symbol, you have a whole lot of possibilities available: at least the ordinary alphabet, the Greek alphabet and a script alphabet. In addition you can use subscripts and superscripts and diacritics.

Do not reinvent the wheel. Try to follow conventions. Speed of light should not be given by "d" or "P", but just by "c". People should be able to guess the meaning at first sight.

7.A.1.A Vectors

Putting arrows over symbols to indicate their character as a vector is old-fashioned. Apply bold instead.

7.A.1.B Diacritics
Every additional accent, subscript, superscript, or bracket decreases the comprehensibility of inline math and displayed equations. Minimize their use.

7.A.1.B.1 Abbreviations as subscript
Math symbols are typically set in italic by default. If your subscript (or superscript) refers to (an abbreviation of) text it should be set in a roman font.

7.A.2 Naming conventions
I already alluded to the fact that international organizations will have naming conventions. Do not reinvent the wheel. Do not use your own idiosyncratic notation.

7.A.3 Units and constants
Use precise conventional notation: the speed of light in vacuum is c, and not c_0. All known symbols in the natural sciences have preferred notations, as determined by international organizations.

7.A.3.A International bodies
International unions in mathematics, physics and chemistry (like *IUPAP*, with its Commission on Symbols, Units, Nomenclature, Atomic Masses & Fundamental Constants, and *IUPAC* with its Nomenclature and Symbols web page, have published preferred notations and units to be used in scientific communication. See also the *NIST Reference*. If your world is still full of ångströms and inches you probably work in a retirement home.

These nomenclature guides are too detailed for everyday use. If you follow the practice found in some of the leading papers in your field, you are probably in the clear.

http://www.iupap.org/
http://www.physics.umanitoba.ca/IUPAP/C2.html
http://www.iupac.org/dhtml_home.html
http://www.iupac.org/general/FAQs/ns.html
http://physics.nist.gov/cuu/Units/

7.A.3.B Dimensions
Variables with a dimension always have a space between number and its dimension. The length is "50 cm" and not "50cm".

7.B Display math (in contrast to inline math)
Mathematical equations, expressions and formulas should be shown in display math as much as possible.

Check that text formatters often indent the next text line after an equation.

You should only allow this if you really want to start a new paragraph there, which is highly unlikely.

7.B.1 Punctuation marks

After an equation a punctuation mark is required: either a period or a comma. A colon after the last word before an equation starts is often redundant.

7.B.2 Equation signs and definition sign

The equation sign is often abused. If your equation is not an equation but a definition, do not use the equation sign "=" but use the definition symbol "≡". You help the reader tremendously with this distinction. If he holds a definition for an equation he might spend a lot of time in trying to understand or rederive it. Whereas a definition is just a handy 'alias' the reader does not need to understand but just needs to remember.

The following symbols are often mixed up:

≈	approximately equal to
~	asymptotically equal to
O(n)	of the order of
→	tends to

The highly-specialized symbol "≅" means "is congruent to" from algebraic number theory. The above prescription may differ from community to community: In the Unicode standard "≃" (U+2243) is defined as "asymptotically equal to".

http://www.unicode.org/

7.B.3 Numbering of equations

Always number all your equations.

There are no exceptions to this rule. It tremendously helps the communication between author and reader, between readers and even between authors, to have all equations numbered. The equation numbers are just navigation anchors.

Do not refer in the text to "this equation", but point to "equation 6" or "expression 7". Only when it is obvious to what equation you are referring to employ "this equation". Enhance and improve your manuscript by referring to an equation not always as "equation (#)" but specify, like: "integral (6)", "approximation (21)", "definition (11)".

Lump equations together if they are tightly connected. If equation (1) refers to the x-direction and equation (2) to y-direction and equation (3) to the z-direction, number them – if allowed by the journal – equation (1a), (1b), (1c). In the text you can refer to them as "equations (1)".

7.B.4 Symbols in equations

Each and every new symbol in an equation needs to be explained shortly before or shortly after the equation.

I repeat, each and every symbol.

If you use conventional notation as much as possible, readers will not need to read your mapping of the symbols but the mappings should nevertheless be there.

7.B.5 Brackets

Brackets inside brackets should differ in type. So no (()), but [()] etc.

7.B.6 Use of appendices

Put difficult math in separate appendices. A short, simple mathematical sideline can be incorporated as a reference (not as a footnote).

7.B.7 Abuse of same symbol

Do not use the same symbol for different things.

Suppose "W" represents a property that has several limits that you want to present explicitly in the manuscript. Give the limiting forms different symbols. So use for instance "W_l" and "W_h" and not "W" for the limiting forms. The use of multi-letter abbreviations is bad practice. So do not introduce "W_{low}" and "W_{high}".

7.C Inline math

Avoid inline math as much as possible. Circumvent defining new symbols inline. Using inline math is playing hide-and-seek with your reader. The ideal is that a colleague should be able to rederive all your equations by merely studying the displayed equations.

7.C.1 Integration with text

Integrate inline math in the text by analyzing what its grammatical function is (it can behave like a noun for instance). Examples: "The density ρ can be calculated", or "For the density $\rho = 4\ \mathrm{cm}^{-18}$, it has been ..."

7.C.2 Beginning of sentence

Do not start a sentence, or clause, with a mathematical symbol.

Rearrange the sentence so that it begins with an alphabetical character. The brains of any reader are trained to interpret a capital as the start of a new sentence. A sentence starting with a symbol will slow down a reader and is just plain repulsive.

7.C.3 Headings with math

Keep away from using math symbols in headings. So not "Calculation of n", but "Calculation of number density".

8 FIGURES

Prepare your figures in such a way that you can immediately (re) use them in your presentations.

Spend a lot of time on producing high-quality figures. I really mean a lot of time. High-quality figures are invitations to read your paper. Try to give the figure a didactic flavor.

A figure, together with its caption, should stand alone and be understandable without the text of the paper.

Realize that editors might pick out one of your figures to feature on the cover.

8.A Default setting

Please realize that the default settings for all options (axes thicknesses, line thicknesses, font sizes of caption, font sizes of labels, data symbol sizes, sizes of tick marks, etc., etc.) are always too small in all commercial programs.

8.B Text aspects of figures

8.B.1 Manuscript text referring to figures

Do not present details of a figure in the running text, but put them in the caption.

In the main text no sentences should appear like "the solid line shows", or "the dashed line represents ..." Just give the message of the figure in the text: "In Fig. 1 we have presented [...] As can be seen in this figure, the agreement between theory and experiment is surprisingly good".

You do not want the reader to continuously look at the figure and come back to the text. The figure-accompanying text and the figure caption are to be read without interruption and without interdependence.

Avoid having additional curves in your figure that are discussed somewhere later in the text. The reader will get confused when he glances over your figure.

8.B.2 Figure captions

Figure captions should always be self-explanatory. Never use in captions sentences like: "See text for explanation". Figures (captions plus graph) should be intelligible without the necessity of reading the text of the manuscript.

Make immediately clear in a caption whether you are presenting experimental data, theoretical data, or numerical data. If your formatter and style file permit, make the caption stand aside from the text. For instance by using larger left and right margins, and by using a bottom margin for your caption. In addition you can use (consistently) a different font for all of your captions.

8.B.2.A Example of bad caption

"Caption to Fig. 2: Same as Figure (1), but with density larger than the threshold density (see text for explanation)." This bad caption is spaghetti-text all the way.

8.B.3 Font size in figures

The default font size for text labels of coordinate axes is always too small in standard commercial programs like *Origin*. Reproduce a figure at the exact size it will appear in the journal. Then show it to a person that is older than 45 without letting him use his reading glasses. Fonts in figures are hardly ever too large.

http://www.originlab.com/

8.B.4 Fonts in figure text (labeling of axes etc.)

You can use whatever font you like under one condition: do not use proprietary fonts.

Some programs, like *Acrobat Distiller* and *PowerPoint*, will not process these proprietary fonts. Proprietary fonts are fonts that you have to pay for to use. I noticed that when I get figure files from some group members these figures sometimes contain proprietary fonts. So pay attention to the type of (default) fonts your drawing or graphing program uses. Use only standard *Type 1* postscript or *True Type*. Font size 10 is unreadable. Minimize the number of different fonts. One font type is usually enough.

8.C Postscript

The standard format for figures is postscript. This file format is a vector format and allows for scaling to any size without loss of resolution. Some journals accept bitmap figures like jpg, gif, or tif. (For your own *PowerPoint* presentations you should always convert your postscript figure files to jpg, to reduce file size and decrease load time of your presentation slides.)

8.C.1 Compatibility

Unfortunately not all postscript is the same. There are a lot of differences between *Unix* environments (much more forgiving) and *Windows* (idiosyncratic) drivers.

> If you want your coauthors to quickly comment on your digital version of your manuscript, test your figures on a *Windows* platform, for instance by using *GSview.*

The newest version of *GSview* is developed by *Ghostgum* (with a delay nag screen if you do not want to pay).

When generating postscript files you have a number of options. Always choose the option with the highest compatibility. For instance by choosing postscript level 2 rather than 3. One way of producing postscript output is to print to a postscript printer driver and force the driver to print to a file. The only thing you have to do in the latter case is to rename that printer file to a file with eps or ps extension.

Problems in postscript files can sometimes be solved by using an ASCII editor and correcting obvious errors in the first few comment lines (starting with "%%") of the postscript (ASCII) file.

http://www.ghostgum.com.au/

8.C.2 Bounding box

> Create figures with almost no blank space around them.

So never incorporate figure numbers, let alone captions, as part of the figure. Such additions make the figures unusable for insertion in other programs.

Figures should display well with *LaTex* style files with only one parameter: the width or the height. If you fail to produce minimal-bounding box figures, others have to edit your figures before they can use them.

Really cool is adding a description of the figure as a comment in your postscript file. This facilitates the use and reuse of the postscript figure tremendously.

8.D Non-postscript

Please do not supply figures as embedded in *MS-Word* or *PowerPoint*. It is not a trivial task to rip these figures out of these files.

If your communication with coauthors is through files which have indeed the figures embedded (like pdf), always supply your coauthors with separate files of your figures in one of the standard formats. Bitmapped figure files are easily converted into postscript (this will not of course increase the resolution), for instance by printing it with a postscript printer driver to a file.

8.E Drawing elements

Create lines, curves and other elements with thick lines and do not vary the line widths too much between the various elements.

8.E.1 Frames

Frames around figures are ugly.

8.E.2 Axes

Always use four axes (so also a right Y-axis and a top X-axis.). In this case the reader gets a much more quantitative impression when glancing over your figure. He can even very easily use a ruler to draw a line in your figure.

Put the position of the right Y-axis exactly at a 'nice, cosmetic' x-value: not "2.95" but "3". Put the top X-axis on a cosmetic Y-value.

8.E.2.A Two different X or Y axes

In some cases there is a choice between two popular scales for the X- (or Y-) axes. A solution is to use the two different scales for the lower and upper X-axis. If possible the upper X-axis should have in this case two types of tick marks: the inner with positions identical to the (inner) tickmarks of the lower X-axis, and the outer according to the new scale (with an appropriate labeling).

8.E.2.B Tick marks

Use large and thick tick marks. You have a choice for putting them inside, outside or both. If possible put them on the inside, unless they clutter up the curves. Inside tick marks guide the eye much better.

8.E.2.C Labels

If the scale on the top X-axis is the same as the bottom put no labeling on the top X-axis (but use identical tick mark positions). The same holds for left and right Y-axes.

8.E.3 Grid lines

Try not to use grid lines, even if they are dashed or dotted. If you nevertheless really have to use them, make the grid lines thin (and/or with a light grey color) and dashed or dotted, so that the curves in the figure keep standing out with a good contrast.

8.E.4 Line thickness

The line widths of curves, axes, and tick marks are always too thin if you use the default settings of your drawing programs. Lines are hardly ever too thick. This is even truer for dotted and dashed lines.

8.E.5 Line style

When more curves have to be plotted using the same X- and Y-axes, the various lines can be distinguished using different line styles: solid, dashed, dotted, etc. Remember that many of the subtle differences between line styles like the difference between dashed and dashed-dotted will be lost when the figure is reduced to standard publication size.

8.E.6 Symbols

If you use symbols to indicate data points, try to always use filled symbols (filled triangles, filled spheres, etc.).

> Filled symbols, in contrast to open symbols, will survive extreme reduction of figure size.

In addition, figures with filled symbols are much better when later you want to reuse your figures for your presentations.

Symbols refer to data points. And data points must have error bars. Period.

8.E.7 Busy figures

Refrain from making busy figures. Avoid figure insets as much as possible. Try not to use arrows to characterize what a curve represents. If you have to use a legend, put it as far away from the curves as possible. Personally, I dislike frames around legends.

8.E.8 Color

Color can make a plot much easier to read. You can color curves and/or symbols and/or error bars. Realize that colored figures might look very different on different monitors. Use forgiving colors and forgiving contrast.

Magenta and orange will always show up ugly or become illegible on a large number of video projectors and printers. Check when applying colors, that, if you print your figure with only gray scales, it is still easy to read. If not, you used the wrong colors.

Having colors in your figures results in many cases to much higher publication and reprint charges.

8.E.8.A Grey scales

Readers might fax your paper to somebody else, or they might print your paper on a grey-scale printer, or, even worse, print it on a black-and-white printer. Take care that your figures are still readable under such conditions (make a red solid line stand out from a black solid line by making it a dashed red line).

8.E.9 Kill dangerous colors

When you use a standard plotting software package like *Origin*, kill the dangerous colors immediately.

For instance in *Origin*:

1. Go to: Format : Color Palette.
2. Now select the unreadable green, cyan and yellow colors and use the buttons on the right to either move the colors down in the *Palette*, or to delete them permanently.

8.F Colorful figures

A paradigmatic plot consist of a left and right Y-axis, a bottom and top X-axis, and one curve. If the data plotted depend on an extra parameter one can plot more curves: each curve relates to a different value of the parameter.

If too many values of the parameter have to be considered, other presentation modes have to be introduced. Commercial plot packages, like *Origin*, allow for various solutions to these situations.

http://www.originlab.com/

8.F.1 Three-dimensional figures

One solution is to resort to a 3D plot. One gets, at first sight, beautiful colorful moon landscapes. Sometimes the plots look more like displaying a collection of Brighton rock candy sticks. At the time of the discovery of the Bose-condensation of cold gasses these phallus symbols were featuring on many covers of physics journals.

I find these candy sticks ugly. A solution is to resort to lighting and shadows in your 3D-plot.

8.F.1.A Professional lighting

If you insist on using a 3D plot you should not make them in many colors.

Only a few colors, or grey scales, and use professional lighting, with shadows. This shadowing helps a lot in getting quantitative information from the plot. *Mathematica* is one of the very few programs that allow for introducing lighting conditions.

http://www.wolfram.com/

8.F.1.B Interpretation

The 3D plots are usually very difficult to interpret quantitatively. They are good to get an overall picture. It gets even worse if the authors – often on purpose – do not show the color scale. Wildly varying colors sometimes only indicate a variation of the magnitude of a variable by less than 10%. You can find these abuses in every issue of *Nature* and *Science*.

8.F.2 Color coding

Rather than plotting one point for a combination of {x,y}-points, one can plot a color on that position.

At the side of the picture a bar with the quantitative color mapping should be given.

Graphs produced in this way look very appealing: a color carpet. Especially in these times where pictures and images are important, scientists easily follow the trend and produce their data in this form. The color mapping is often 'forgotten' as it would spoil the party.

From the negative side there are several remarks to be made. It is very difficult to look at the graph and get quantitative information without studying it at length in great detail.

I think this way of presentation is only acceptable as the color mapping is shown explicitly (leaving out the color mapping is immoral). In addition one should present some cuts through the color graph.

8.F.3 Cross-sections

In any case, when one presents data in 3D form and/or data with color maps, then a few cuts, resulting in old-fashioned X-Y plots should also be presented.

Comparing a 3D color plot obtained from a theory with a 3D color plot from experiment is usually done to hide the poor agreement between the two. A scientist with guts dares to show conventional graphs (cuts though the 3D-color carpet) comparing theory, without using arbitrary units, to experiment.

9 TABLES

Tables, together with their caption and title, should constitute an independent piece of information that can be appreciated without reading the main text.

Tables should be uncluttered. Avoid using text entries. So no "na" for "not applicable", just leave the entry blank (no dash either). Use the last or first column, or a footnote to the table to make clear why there is no entry.

9.A Columns and rows

If you have your table prepared, check whether you can switch columns and rows.

If so: choose the arrangement where the comparison between similar entries can be done vertically, and not horizontally.

Human brains are much quicker in comparing vertically aligned numbers, than equally-spaced horizontal numbers.

9.B Justification

Never use left or right alignment for column entries representing numbers. Use a form of decimal alignment of the entries: vertically align them on the decimal point if present, otherwise vertically align them with their least-significant digit vertically aligned. Avoid using exponential notation for numbers. Following these rules will result in tables for which only a superficial glance gives a reader already a good impression of where the large and where the small numbers are.

9.C Caption and title

Use a very smart caption and short title that together form a self-contained unit.

9.D Reuse

Try to make columns as narrow as possible, for instance by using two lines for the text headers.

You might want to use the table later for your presentations. This recycling requirement puts large constraints on the maximum number of columns and rows that you can introduce.

10 SUBMISSION

10.A Permission

Submission of a paper requires permission from a senior member of the group. Your institute might have more formal procedures. The reason behind this bureaucracy is that the content and quality of the paper reflects back on the whole group, and on the whole institute. This connection has moral and legal sides. Not following these procedures can have very serious consequences. If you want to send in a paper at your own responsibility only, use your private home address as affiliation and refrain from using any facilities of your employer.

10.B Submission letter

> Never put scientific arguments in the submission letter to the editor.

Reasons for publication should be in the paper. You could describe in four or five lines the important conclusions of your work for non-specialists. Many journals request this popularized summary explicitly.

> Give a list of (at least three) possible referees (with full affiliation, email, website).

Add a sentence that these referees have not been contacted by any of the authors, and have not been informed in any other way about the content of this paper.

Asking the editor to exclude a referee, who you know is hostile towards your work or person, is dangerous. The editor might be tempted to use him out of all people. If you want to rule out a hostile referee, explain to the editor why you want to bar him.

Rather than excluding a hostile referee, a better way is to cite a number of his papers in a positive way in your manuscript.

Few people can resist the temptation of a boosted citation record.

10.B.1 *Nature* and *Science*

A number of prestigious journals, like *Nature* and *Science* subject the manuscript to a real test before they send it out to referees. I hate this but I am not a Don Quixote. So I also submit manuscripts to them.

You can influence this process in a number of ways. Firstly in the manuscript itself. But also in the correspondence accompanying the submission process.

10.B.1.A Manuscript

Include in your list of references a few referrals to work published recently in the same (type of) journal(s) you are going to submit your manuscript to.

Do not overdo this, because then the editorial office could use as rejection argument that the subject gets too much coverage already.

In the acknowledgement (possibly as a reference) thank one or two heavyweights for their enthusiastic support and their insightful comments. A private communication to a Nobel-prize-stature scientist will certainly help.

A number of (divisional) editors of the magazines write books and feature articles in other journals. Refer to them in your manuscript and your supporting letter.

10.B.1.A.1 Good-news show

Whatever the content of your paper is, communicate it in the form of a good-news show.

The editors of the prestigious journals hate bad news and love good news (unless you predict Apocalypse Now due to global warming). Irrespective of the scientific basis for the claims. Part of their reasons of existence is creating good news. So if your message is a very original, well-founded, critical assessment of a hyped-up field, forget it. It will never get published in the magazines. You really need to go to the professional journals.

If your paper is the fifth or seventh paper supporting outrageous claims of an, in the eyes of reporters, promising field, you stand a big chance.

There is a famous reply by an American scientist when asked what he does when his audience criticizes him of not having delivered any of his previous promises. His answer: "I just promise them more".

10.B.1.A.2 Math

Minimize the math in your paper. Any additional equation will half your chances of being admitted to a round of referees and will in addition half your chance of being finally accepted.

10.B.1.A.3 Graphics

Minimize the number of dull X-Y graphs. Put real colorful pictures in the manuscript. Whatever graphic material you can use, use it. Do it with colors and certainly with 3D plots. I know that these plots are horrible, but it is what the journal wants.

10.B.1.B Submission letter

As the editors claim to be able to judge on the basis of science, you should put scientific arguments in the submission letter. Apply some name dropping of famous scientists that endorse your results. Refer to editorials or other magazine feature articles.

10.C Mode of transport

Submit by email/internet if possible. If submitted by mail, always use a courier service (it will at least impress the editor).

11 REFEREE REPORTS

The communication with the editor and referees is a matter that concerns all authors. Independent of who writes the replies to referees and editors, the corresponding author should always have the consent of all authors before sending in replies. Any new correspondence obtained from the editor should immediately be forwarded to all coauthors.

11.A Answering reports

Answering referee reports is a crucial aspect of getting your paper published.

> Rule number one is: "The referee is always right." Rule number two: "In the exceptional case that the referee is wrong, do not tell him, and do not tell the editor."

Try to get them all on your side. Whatever it takes.

11.A.1 Who will write the replies?

The replies to editor and referees should be written by the first author. He should be supported by a senior coauthor with lots of experience. The other authors should keep a low profile as the timing of the replies is crucially important: try to return the replies to the editor within one or two days after having received the referee reports. This quick response will impress editor and referees. Of course all authors should be kept informed about the progress of the reply process.

11.A.2 Styled text

The correspondence between editor/referees on one side and the authors on the other should be through ASCII files (usually in the form of email appendices or email attachments). No styled documents, with font formatting etc. So replies to referees should be made with an ASCII editor. Communication between referee and editor very likely goes through email or web access. In all those cases styling will be lost.

11.A.3 Style of rebuttal

Always be very, very polite. Non-American authors (especially Dutch, Israeli and Russian) tend to be too rude in their replies. A little flattering of the referee is appropriate.

Sympathize with the referee: He has read your paper and has written a report, all without any financial compensation. He would have preferred to get on with his own work or go sailing. Your final goal is to get your paper published. Never let your ego get in the way.

Be very careful with humor, from gentle to sarcasm. With a non-native English speaking author and/or a non-native speaking referee this humor is bound to be taken wrongly.

Think about the French waiter of a first class restaurant serving an American couple with no knowledge whatsoever of the French cuisine. By staying patient, humble, friendly and inexhaustible he is superior to the couple.

So be very humble to the referee. Show, without exaggeration, that you very much appreciate his efforts. The last sentence of each rebuttal should read something like "by implementing the very useful comments of the referee we hope that he will support publication of the much improved manuscript".

11.A.4 Header of replies

The reply to the referee should contain a header on each page with information about authors, title of manuscript and journal.

So if the reply gets temporarily mislaid or gets on the floor it can be placed in context without the help of *Crime Scene Investigations*.

11.A.5 Opening remarks rebuttal

Take immediately the sting out of a referee report.

After you open your reply with acknowledging his efforts, you come with a very positive summarizing sentence:

"We are glad that the referee finds our results very interesting", or "We are pleased that the referee considers our experiments of high quality and our results well-presented." Even if the rest of the report was full of negative criticism.

11.A.6 Scientific conflicts

If the referee comes with stupid scientific arguments, do not tell him. Just tell him that because of your inadequate presentation the referee could not help to be misled, and that the question will become clear in the new manuscript as you have improved the presentation. And do change the presentation and do tell the referee that you changed the presentation of your paper because of his useful comment (which in essence was just a mistake on his side).

11.A.6.A Referee finds a flaw

If the referee finds an important mistake in your paper you have a real problem. If the error can be corrected and would change the manuscript only in a minor way: be very humble, change, and resubmit with acknowledgement.

If the mistake is substantial, withdraw the paper. Not telling the editor and referees explicitly that you withdraw the paper is cowardly.

11.A.7 Length of rebuttal

Rebuttals should be short. Never give scientific arguments in the reply that should be in the paper. If the referee makes a scientific argument that is wrong, do not tell him by presenting a derivation in your rebuttal. If his mistake is a plausible one you should put the scientific argument in the paper.

Never use long paragraphs. Lengthy arguments give the impression that important information is missing in the paper.

11.A.8 Manuscript modifications

Address all comments of the referee.

> Begin each reply to a particular comment always with summarizing his comment in a slightly more positive way than intended by the referee.

The referee will not feel the necessity to consult his earlier report and you have already gained some ground.

Always make (small) modifications on the basis of each and every comment raised by the referee. Tell the referee in detail you did this because of his report. In this way the referee starts to feel responsible for your text.

11.A.8.A One referee at a time

Never tell a referee in a reply about the presence of other referees. Never use the following example when rebutting Referee B: "In contrast to Referee A this referee disagrees with ..." In your rebuttal to Referee B there is only one world: you and Referee B. And nobody else. Give the referee the impression that you take him very seriously. Make him feel important (which is what he is).

Unfortunately you cannot avoid that the editor supplies each referee with the reports of all referees. I consider this bad practice and laziness on the side of the editor. Write your referee rebuttals such, that if this occurs no accidents occur.

11.A.8.B Modifications due to other referee

An unfortunate circumstance occurs if more than one referee, let say Referee A and Referee B require substantial modifications. Sell the modifications that Referee A requires in a subtle way to Referee B. Explain in the rebuttal that the authors found this cosmetic change necessary to improve the presentation.

11.A.9 Conflicting referee reports

If you have referees with strongly conflicting reports requesting major changes, you have a real problem. Neither of the two will ever give in. The editor will never take sides. Try to find contradictions in each individual report.

As an emergency exit ask for an additional referee that will have access to all reports. Rebut all reports, even if they will not be sent to the referees.

11.A.10 Additional references

It often occurs that referees ask in their report to include new references. You can bet that the referee is either a coauthor or he is closely related to these authors. Give in.

If you are sure that the new paper you have to include is not correct, or of poor quality, put it in your list of references anyway. Cite it in a neutral way in the paper (embed it in a long list of citations). Or find a relevant, good-quality paper of the same authors and cite that paper.

11.B Perseverance

Never give up on a referee. Try to get him on your side. This might seem hopeless, but it will impress the editor and later referees that will see the whole package of referee reports and replies. Even if the editor tells you he will not consult Referee A any longer, still write a rebuttal to Referee A. The editor might review his decision, and later referees will see this rebuttal. The idea is that there should not be any scientific criticism raised by any referee that has not been answered by you.

11.C Letter to the editor

The letter to the editor accompanying your rebuttals should be very short. Always refer to editorial assistant, assistant editor, etc. as "Dr. Last Name" and "Editor". You can be sure that their ambition will be to become a full editor, and getting a PhD degree if they do not have one yet.

The letter to the editor should be about procedures. Never about scientific content.

If you raise a scientific question in this letter, the editor might want to consult a new referee, which would just mean a delay.

11.D Resubmission package

Send in your replies to each referee and your letter to the editor as separate files. If the journal policy forbids this, make it very easy for the editor to separate the individual items into files (individual reports and letter to the editor).

The worst thing that can happen to you (unfortunately this happens very often) is that each referee gets the whole package (letter to editor plus all referee reports and all replies). The problem is that negative comments of Referee A tend to have a large negative effect on Referee B and vice versa. So use \newpage if you submit as *LaTex*, or use a hard page break if you use *MS-Word*, or use other separators. If you are desperate, you can violate the journal policy and explicitly send in separate files.

When you resubmit, tell the editor you want the individual referees only to receive the correspondence you specially prepared for them.

11.E Failed submission

In almost all situations editors perform a marginal check on a submitted manuscript. A number of trivial points are verified: presence of title, of list of authors, of abstract, of content, and of references. Absence of many grammatical and spelling errors is another secondary criterion.

If a manuscript passes these marginal tests it will be sent to one or more reviewers.

However a few number of prestigious magazines, notably *Nature* and *Science,* have their own divisional editors actually read the paper and reject a large number of papers on the basis of a number of very unclear criteria. Those papers will never be sent out for review.

I strongly disagree with this lottery procedure. The problem is that divisional editors, their assistants or the editor himself, are non-active researchers. No matter how experienced they are, and no matter how well-informed they are, and no matter how good they are as journalists, they are not active researchers. These non-experts make expert judgment calls. They are bound to make mistakes, large in size and large in number. The good thing about this, in my opinion, highly questionable procedure is of course that the journal needs an order of magnitude less referees.

11.E.1 Protest against refusal of review

A refusal by a (divisional) editor of *Nature* or *Science* to subject your paper to peer review is close to being a fatal rejection.

However, if you are convinced that your paper should be sent out for review, you should appeal to the rejection by the non-expert (divisional) editor.

In principle you should be furious because you should only submit your very best papers to these prestigious journals.

Analyze the arguments of the magazine editors carefully. Try to find an inconsistency or an unfairness in them. Check a large number of papers that the journal did accept, preferably on the same subject. Confront the editor with their unpredictable, erratic behavior.

Make your protest heavier by getting heavy-weights on your side (explicitly in your protest letter) or by improving the acknowledgement part of the newly submitted manuscript. Call the responsible editor on the phone. They are in many cases just people that left science at the level of a postdoc.

If you submit a large fraction of your manuscripts to these prestigious magazines, the editors realize that you abuse their submission process. In such a situation you should not protest. You just suffer from a too large ego.

Even if you do not win the fight, they will remember you forever. If you lost the fight, tell them that you will submit somewhere else. Tell them that they are not allowed to use any of the information in the paper or in the submission letter for their journal (Trust me: do it). Submit your paper elsewhere. When it gets published, try to get as much publicity as you can for your results. And confront the hostile journalist-editors with the success of your scientific paper, a paper they rejected.

11.F Writing referee reports

Reviewing a manuscript is time consuming. You do not get any money for it, but it is part of your social duty. If you do not want to be a referee you should not expect other people to review your papers.

If a referee finds that the paper he reviews violates a number of hints presented in this *Writing Guide*, he should point this out, in detail, in his report.

11.F.1 Pride

Young researchers are often pleased to find that renowned journals ask their advice. That is good, but they should be careful: before they know it they are buried under reviewing jobs. Especially when a referee does a good job, editors will consult him over and over again.

11.F.2 Reviewing load

Try to minimize your load. One golden rule is: submit your review report (way) over time.

If you submit your report way before the deadline you will be the favorite of the editor with an ever increasing load.

Do not review for journals you yourself would never publish in.

When you refuse to review a paper explain in some detail to the editor why. "I understand that reviewing is important [...] However, the number of papers I have to review is overwhelming. To survive as a scientist I have to decline [...]".

11.F.3 Alternative referees

Have a list of alternative reviewers with their expertise available.

Do not fall into the trap of passing on the manuscript yourself to your colleague, postdoc or PhD student. Tell the editor who is your favorite choice for replacing you and ask the editor to contact that person directly – even if the guy is sharing an office with you. If you do not follow this advice you will remain partly responsible for the reviewing process and you will get the reminders if your colleague is falling behind.

11.F.4 Standard format

You are doing the journal a favor (whatever the editors of *Nature* and *Science* want you to believe) when you write a referee report.

The quickest way to write a report is to use your own format in plain ASCII (fixed pitch font).

If the journal does not accept that, too bad for them. Here is an example (use abbreviations, use et al., and no initials for author):

REFEREE REPORT

Title: The observation of ...
Authors: Elian et al.
Journal: J. Chem. Phys.
Manuscript: 5678

GENERAL

The paper describes an interesting [...]

The experiments seem sound. However the paper is not well-written. The equations are quite sloppy. Following my conclusion I will present my detailed comments.

CONCLUSION

In its present form the manuscript should not be published. I advise the authors to submit a revised version taking into account my detailed comments. If all of my objections are met I will probably advice the editor to accept the modified paper.

DETAILED COMMENTS

1 In the abstract

The abstract promises something that is not actually presented in the paper.

2. on page 3 "the theory"

This statement is wrong ...

3. Eq. (12)

4.

[...]

11.F.5 Quality of manuscript

Of course I cannot comment here on the scientific quality of a paper you are reviewing. However, it is also the task of the referee to check the *presentation* quality. If the referee finds that some important hints of the *Writing Guide* advise are flagrantly violated, he should refuse the paper. I refer to items as quality of figures, captions, tables, partitioning of the paper, conclusion, etc.

12 ADMINISTRATION

Details (full author list, title, journal, etc.) of the submitted paper should be sent to a secretary-administrator of your group or institute. He should also be informed about revision submissions, changes of journal, retraction, acceptance. The more people know about your paper, the higher the chance it will feature in institute reports, yearly reports, etc.

12.A Publication costs

In principle always pay publication charges if you are in a rich country.

Nowadays journals can ask you money for making your paper available on-line, free for any reader. Pay. The amount of money is negligible compared to the amount of money that the research described in your manuscript has cost.

12.B Reprint orders

Reprint orders are always handled by a secretary-administrator. He will also manage the reprint database. Do not keep a large stack of reprints for yourself. Ten copies are more than enough to impress your mother, your girl friend, and other close people.

Nowadays many institutes refrain from ordering hard-copy reprints. The combination of a pdf file and a high-quality local printer is supposed to be sufficient.

12.C Mailing list

Consult with your coauthors asap about the mailing list.

This could influence the number of reprints being ordered. Nowadays the success of the internet has made the use of mailing lists less important. Nevertheless, sending your colleagues regularly a stack of recent papers is not a bad idea.

12.D Digital Archive

12.D.1 Group file server

It is very likely that your group has a file server where all group members can read and place files.

Each group member should have his own directory on this server. Apart from a possible private subdirectory, all other information should be readable by other group members.

Check that your group members can read but not change files in your subdirectory. Give your subdirectories and all your files sensible names, that is to say: names that make sense to an English-speaking person. And please: no spaces (or dashes) in any of those names. Use one type of case (lower) only and use underscores as separators. In this way your file names will be (computer operating system) platform independent.

You do not own the information in your own directory. So when you leave the group, absolutely do not erase that information. Ask your group leader to freeze your directory, but keep it available to present and new group members.

12.D.2 Saving files

Together with your submission, you should put an exact copy of all the files sent to the journal on the group file server. Include also a copy of the submission letter, etc. The directories should also contain referee reports that will be received on the present version. If you receive correspondence by a journal that is not in digital form, scan that information and put the scanned images there (no OCR please)

All coauthors should be immediately informed about the location of these files. If possible use a flat directory: that is one directory per version of the paper (no figure subdirectories). In such a way the whole contents of the directory can easily be zipped and if necessary, mailed or resubmitted. Use the submission date as directory name: So an example would be

//bret/articles/gap_localization/2007_11_04

Name the last directories in such a way that alphabetic listing by explorer type file managers shows them in chronological order (year, month, day, with leading zeroes). Do not use spaces in a directory/file name. Nowadays hard disk storage is so cheap that there is no need to delete the old submitted versions.

It is useful to give the directory that contains the accepted version a name that signifies that it contains the accepted paper.

This renaming should be done such that alphabetical listing still produces the chronological order (use underscores for instance to force the right alphabetical listing).

The whole idea behind the archiving discipline is to increase tremendously the chance that a group member will use your figures in one of his presentations. So it is also very good for your own career. Trust them: they will give proper credit to you if they use it.

Furthermore, you make life much less hectic for your group members. Your senior group members are bound to be asked to supply recommendation letters in which they have to evaluate your functioning. Make life easy for them. I often have to copy a figure out of a group-member directory in the middle of a Sunday night when my plane leaves on Monday morning. Adopting the here prescribed rules helps a lot.

13 REACHING OUT

You must have anticipated the possible acceptance of your paper. In some cases (20%) you will find your paper very important (already reflected by the journal you have chosen to publish it in). In that case read the next item.

13.A Publicizing your work

Try to get your work in the media. A number of professional organizations have special news sections in their glossy magazines and on their websites.

> The best way to get your paper in there is to check the format they use. Write exactly in that format in a very popular way. Have it checked by non-specialized colleagues. Send it in. Do not ask media representatives first if they are interested.

In contacts with reporters try to avoid the service of spokesmen of your institute.

Write a press-release yourself. Have it read by very junior scientists and by a non-scientist. Put any success you have with the media on your website and mention it in your annual reports and grant proposals.

Be sure that all your managers will get to know about your media success.

14 ALTERNATIVE PUBLISHING

14.A Official: preprint server

The review process of your paper might take a long time with no guaranteed success. To establish your scientific priority immediately, you can submit your paper in addition to an established preprint server.

Realize that some journals, including *Nature* and *Science*, do not allow this parallel publication channel.

In exceptional cases (<10%) you might want to put off the point in time that your colleagues can inspect your results to the latest possible moment. In such a case you should not submit to the digital archive. In all other cases inclusion in the database is beneficial.

14.B Unofficial publishing

There are a number of scientific manuscripts that will not be published in an international journal, among which: (undergraduate and graduate) theses, grant proposals, and internal reports.

Even if your paper will be published, you might be inclined to supply colleagues with other possibilities, like web posting, of obtaining your paper.

14.B.1 Format

For these unofficial papers you are the master of the format and free in the choice of the formatter. The formatter will either be a member of the *LaTex* family, *MS Word* or the like, or an html editor.

In the majority of cases the final form of your paper will be in pdf or in html format.

> Put a copy of the web-posting file(s) also in your folder on the group server.

14.B.1.A Pdf files

Nowadays making high-quality pdf files is a piece of cake with the *Tex*-family. If you use *dvips*, just use the *-Ppdf* option. Making a pdf file out of an *MS Word* document is best done with *Adobe PDFMaker*, a very buggy, but very useful software utility supplied with *Adobe's Acrobat.* Unfortunately starting from version *7 Adobe* does not supply this piece of software anymore. The new *Adobe Pdf printer driver* is much less powerful. So please, if you upgrade, keep an old version of *Acrobat* as well. In *MS Office 2007* files can finally be exported as pdf.

14.B.1.A.1 Pdf fonts

When using a *Windows postscript* driver, avoid using *Type 3* fonts by preventing the driver from sending the fonts as bitmaps. The pdf file should only contain *Type 1* and *TrueType*. Please check the absence of *Type* 3 fonts before distributing your pdf file (*File, Document Properties, Fonts*). Consult an article from the *Adobe* knowledge base for details.

http://www.stringcat.com/guides/writing/files/whitehouse.htm

14.B.1.A.2 Layout of paper

> For screen viewing two-column single-space text is cool (two-column text has shorter sentences, helping the eyes to read the text much quicker).

With *MS Word* this reformatting (with mathematical equations) into two columns is murder. Forget it.

With *Latex*, creating two columns is quite simple. Many macro packages, the majority of them developed by journal publishers, do the job. Avoid the very confusing alternation of partly double column and partly single column text (to fit wide pictures or equations).

If you are able to format your manuscript as to be displayed in a two-column format, do it by all means.

14.B.1.A.3 Figures in pdf files

Often journal editors do not allow you to position your figures in your main text file. They usually want each figure on a separate page at the end of the manuscript, either physically there or as an inclusion directive. But the resulting pdf file – with all figures at the end – is unacceptable for web posting.

> So if you want your paper on the web as soon as possible, and you want it as a pdf file, prepare an additional version of your main text source file that does insert the figures in the right place, which is close to where they are referred to first in the text.

Avoid putting or editing the figures directly into the pdf file. If you do so anyway you will have to redo this manual insertion/edit again everytime you change the source file.

14.B.1.A.4 Figures in single-column pdf files

With *MS Word*, positioning figures in single-column pdf files (preprint format) anywhere in the manuscript is quite simple.

With *Latex*, positioning of figures is possible but not easy. Do not check any of the many cumbersome online manuals, but check an example of a *LaTex* source file (from one of your more senior colleagues for instance) where this insertion of figures has been done successfully.

Latex likes to put figures at the top or bottom of a page (or column). You will have to go through a number of compilation cycles of adjusting height/width, insertion of vertical space, and relocation of the insert directive before the figure will be displayed at an acceptable position in the manuscript.

Put a copy of this web-posting *LaTex* file in your folder on the group server also.

14.B.1.A.5 Figures in two-column pdf files

In a two-column manuscript figure insertion with *LaTex* is even trickier. Figures should be of high quality as they will be reduced considerably. Try to have as little as possible surrounding white space in your figure. The rest is a matter of trial and error. But in the end you will be able to get your figure where you like it.

Put a copy of this web-posting *LaTex* file in your folder on the group server also.

14.B.1.A.6 Navigation

Scientific texts, either in pdf or html format, of which the format is not prescribed by a journal, should have extensive navigation tools. These tools are bookmarks in pdf and hyperlinks in html.

The bookmarks in pdf can be generated from some versions of *MS-Word* or with the *hyperref* package in *LaTex* (here is a hyperref tutorial).

http://www.tug.org/applications/hyperref/manual.html

14.B.1.B Postscript

Papers formatted as postscript should not be used.

Gzipping them is even worse. Pdf is the standard. Postscript viewers are unwieldy and have very poor screen quality.

An additional drawback is the existence of many flavors and versions of postscript, giving rise to a number of – sometimes very subtle, but very irritating – software and printer incompatibilities.

14.B.1.C Web page

If you publish your paper on the web, there might be a number of reasons why you would want to present your paper formatted as html (in addition of course to supplying a pdf version). You are in good company: many professional journals, for instance *Nature* and *Science* supply html versions of their contents.

14.B.1.C.1 Html formatting

If your paper has been prepared with standard word processors as *MS-Word* or the like, just use *Save As Web Page (filtered html).* You should not have embedded your figures in the *MS-Word* file, but they should be in there as links. You really want to clean up the non-standard html *MS-Word* has produced, although you did save them as 'filtered' html. Cleaning up your web pages with an html editor like Dreamweaver works like a charm.

http://www.adobe.com/products/dreamweaver/

14.B.1.C.2 Navigation

If you want to improve the navigation through your (long) web paper, readers expect a lot of hyperlinks. They expect a navigation pane with a collapsible tree.

Generating a navigation tree in html from an *MS-Word* or pdf source is best done with commercial software (for instance *Art HTML Listing* from *ZZEE*).

http://www.zzee.com/

14.B.1.D Desktop publishing

If you have to produce, or just want to produce, fully professional, fully justified text and the *Tex* family cannot be used, for instance because your source texts are delivered in *MS-Word* format, you have to step up.

You need desktop publishing software like *Adobe InDesign* (quite expensive). And you only need a fraction of its functionality: micro-adjustments of word and character spaces. Output can be printed or exported to a pdf file with full retention of quality. Inexperienced users should expect to be confronted with a learning curve of at least a full week before the first pdf file can be produced. The presence of a number of bugs requires smart bypassing.

14.B.2 Copy right and web posting

You might wonder whether or not you are allowed to put your own published papers on the web. Some commercial publishers bark at you when you try to do so. Just call their bluff. You can put on your website copies of your own work, even if these are exact copies of the commercial publisher files. Do not tamper with these files. Do not remove copyright texts (you can remove irritating water marks). As long as it is your own work, and as long as its use is non-commercial, there is no problem. Whatever the publishers say.

In the USA a number of website owners have been threatened under the Digital Millennium Copyright Act. However, I know of no case where individual scientists have been threatened by publishers. I still think you should call their bluff.

15 PROTECTING YOUR PAPERS

There are a number of reasons to protect the content of your papers. Possible motives are commercial interest, or protecting first discovery claim.

15.A Reasons

15.A.1 Commercial reasons

You want to make money out of your paper and do not want unauthorized copying. For scientists these commercial reasons are usually not relevant. It hardly ever occurs that a scientist makes money on a scientific paper. It is indeed against the culture of science.

15.A.2 First discovery

If your manuscript gets published in an official journal or accepted by a digital archive, your first-discovery claim is assured.

But what if, for instance, you only want to publish it on a website? This could be relevant for didactic papers, tutorials, lecture notes, and presentations. You might want to prevent people from copying your site. You want to ward off evil colleagues to republish your manuscript under their own name.

How do you avoid abuse of your unofficially published material?

15.B Protection solutions

15.B.1 Regular publications

If you publish your paper in a regular journal, that itself is the protection. In that case reuse and abuse of your text and figures can not be prevented. Any scientist can go to the website of your publisher, copy your figures and use them. Publishers are very bad at protecting your files. It is not really in their interest.

15.B.2 Watermarks

Watermark protection is weak, but it warns innocent users who are about to become abusers. A hacker will be able to remove any watermark, but it does require some expertise.

You can make life more difficult for a hacker by merging watermarks as a figure with the background.

As an illustration: you can superimpose a very large, almost transparent, diagonal text with "NOT TO BE COPIED: <YOUR NAME>". Merge the text as a figure. Then try yourself to remove the watermark with the reading/editing program you expect the abuser will use (for instance *Adobe Acrobat*). This will tell you how difficult it is.

15.B.3 *MS-Word* documents

Papers being distributed as *MS-Word* files cannot be protected. Period. All the password stuff in it is just frills. Any evil hacker can simple copy your paper and change it.

15.B.4 Protecting pdf files

Pdf files can be protected in several ways.

15.B.4.A Commercial copy protection

There are a number of solutions:

Reading of your pdf files could be implemented as to require an extra plug-in in *Adobe Reader* (for example *FileOpen's WebPublisher*).

Programs that wrap a pdf file in a wrapper and make an exe file out of it. Exe files can be much better protected. Proprietary viewers unwrap the wrapper and show the pdf file. A big problem is that bookmarks never survive this wrapping. And of course it is not platform independent.

Other solutions change the pdf file into a different file format that requires a proprietary viewer.

http://www.fileopen.com/

15.B.4.B Access protection

Adobe Acrobat allows a high level of protection by encryption on its pdf file. You can disallow reading, changing, selecting, printing, etc. All protected with a password. If you however do not use the user password (if you do not disable reading without a password) a number of cheap commercial hacker programs allow evil readers to crack the other protections.

Effectively the best protection is to distribute your pdf files for free. As there is no commercial value in the paper no hacker will spend time on this.

15.B.4.C Disable text copying once and for all

If you are not afraid of large pdf files and you do not need scalable (use of zoom facilities) pdf files, you can make life for hackers unpleasant in another way.

Export your pdf file as jpg pictures (with possible inclusion of watermarks). *Acrobat* will produce a picture file per page. Make a new pdf file by importing all these pictures all at once.

Go back to your original pdf file, with the bookmark navigation tree, and replace all pages (with text contents) all at once with all the picture pages. You will now have a large pdf file of reasonable quality with good bookmark navigation quality (hyperlinking in the text will be lost). As there is no underlying text anymore, abuse of your results is much less attractive for abusers.

15.B.5 Websites

Website protection seems impossible, but you can make life difficult for hackers.

Pages can be encrypted and protected against a lot of copying.

(Example: the commercial website protection software *HTML Protector*). If, in addition, you use only absolute website hyperlinks and no relative links, off-line reading will be very cumbersome.

http://www.antssoft.com/htmlprotector/index.htm

16 ABOUT

16.A Abbreviations

Abbreviations not explicitly explained in the text are:

ACS	American Chemical Society
AIP	American Institute of Physics
APS	American Physical Society
IEEE	Institute of Electrical and Electronics Engineers
IUPAC	International Union of Pure and Applied Chemistry
IUPAP	International Union of Pure and Applied Physics
NIST	National Institute of Standards and Technology
OSA	Optical Society of America

http://portal.acs.org/portal/acs/corg/content
http://www.aip.org/
http://www.aps.org/
http://www.ieee.org/portal/site
http://www.iupac.org/dhtml_home.html
http://www.iupap.org/
http://www.nist.gov/
http://www.osa.org/

16.B Trademarks

Acrobat, Adobe Reader, InDesign, and Distiller are registered trademarks of *Adobe Systems Inc.*
http://www.adobe.com/
WinEdt is a product of Aleksander Simonic (alex@winedt.com).
http://www.winedt.com
HTML Protector is a product of *AntsSoft.*
http://www.antssoft.com/index.htm
Macintosh is a registered trademark of *Apple Computer Inc.*
http://www.apple.com/
MathType is a product of *Design Science.*
http://www.dessci.com/en/
ScienceDirect is a registered trademark of *Elsevier B.V.*
http://www.sciencedirect.com/

FileOpen WebPublisher is a registered trademark of *FileOpen Systems Inc.*
http://www.fileopen.com/
GSview is a product of *Ghostgum Software Pty Ltd.*
http://www.ghostgum.com.au/
Google is registered trademark of *Google Inc.*
http://www.google.com/corporate/index.html
Thesaurus.com is a product of *Lexico Publishing Group, LLC.*
http://www.lexico.com/
Scientific Word is a product *MacKichan Software Inc.*
http://www.mackichan.com/
Dreamweaver is a registered trademark of *Adobe Systems Inc.*
http://www.adobe.com

CD-ROM Emulator is a product of *Paragon Software Group.*
http://www.paragon-online.de/
ISI Web of Science is a registered trademark product of *The Thomson Corporation.*
http://www.thomson.com/
Mathematica is a registered trademark of *Wolfram Research, Inc.*
http://www.wolfram.com/
Microsoft Word, PowerPoint, and Windows are registered trademarks of *Microsoft Corporation.*
http://www.microsoft.com/en/us/default.aspx
Origin is a product of *OriginLab Corporation.*
http://www.originlab.com/
Merriam Webster is registered trademark of *Merriam Webster Inc.*
http://www.merriam-webster.com/
Art HTML Listing is a product of *ZZEE.*
http://www.zzee.com/

PRESENTATION GUIDE FOR SCIENTISTS

1 GENERAL

The *Presentation Guide for Scientists* (from now on: *Presentation Guide*) you are browsing (or even studying) at this moment is part of the *Survival Guide For Scientists* (from now on: the *Survival Guide*). Occasionally there might be referrals in the *Presentation Guide* to other parts of the *Survival Guide.* However, the *Presentation Guide* is supposed to be self-contained. For completeness we present here the names of all the guides that together constitute the *Survival Guide*:

- *Presentation Guide for Scientists,* or short: *Presentation Guide*; as an addendum to the *Presentation Guide* we have published the *Example Guide*
- *Email Guide for Scientists,* or short: *Email Guide*
- *Writing Guide for Scientists,* or short: *Writing Guide*
- *Survival Guide for Junior Scientists,* or short: *Junior Guide*
- *Survival Guide for Senior Scientists,* or short: *Senior Guide*

The *Presentation Guide* lays out a set of rules that are helpful when presenting scientific information in front of an audience. The compilation of rules described in the *Presentation Guide* is supposed to form a reference text with easy navigation.

To avoid having to duplicate too much from the *Writing Guide,* some parts of the *Writing Guide* will be referred to and not reproduced here.

These references to the *Writing Guide* include references to the detailed guide lines about how to make high-quality figures and tables. Furthermore only a few of the many spelling hints will be reproduced here.

1.A Target group

The target group I had originally in mind was physics undergraduate and graduate students, and physics postdocs. From experience I have discovered that many of the more senior physicists (90%) could profit from studying the set of instructions laid out in the *Presentation Guide.*

Studying the principles will not only help physicists to deliver better presentations. The text is very likely also beneficial to mathematicians and to workers in other natural science disciplines, like chemistry and astronomy.

The content is highly modular. The number of cross-links in the guide is kept to a minimum. Each item can be studied on its own. Researchers in other fields can easily skip parts they consider too closely related to physics, or which they deem irrelevant for other reasons.

But it doesn't stop there: many of the hints are of invaluable use for a much broader audience of professionals.

1.B Goal of presentations in general

Whenever I listen to a talk, I am trying to find out what goal the presenter has in mind. After having listened to many (more than 500) talks I come to a classification that will be presented in the next item.

1.B.1 Classification of talks

I have arrived at the following categorization:
a) speaker has no goal apparently
b) speaker wants to prove he is smarter than anybody in the audience
c) speaker wants to give listeners a flavor of new developments in his field
d) speaker wants to teach some new science to the listeners.

1.B.2 Talks of type a)

There is not much I can say about this type of talks.

1.B.3 Talks of type b)

The disappointing feature of lecturing is that talks of type b) often are very much appreciated by the audience. The speaker comes with chaotic – but intimidating – slides, in false colors and with a lot of impressive formulas. He utters a lot of one-liners that are popular in the media. The orator speaks fast to extremely fast. There is no chance for the listeners to understand anything. The speaker gives two or three talks rather than one. He goes way over his time. He is abusing the question period to deliver the remaining part of his presentation. Any real question is received as an insult. The prevailing opinion in the audience is: "This guy is extremely smart and very enthusiastic. I wish I was that smart."

I hate presentations of this kind. Giving talks of type b) is very easy. Too easy for a real scientist.

1.B.4 Talks of type d)

Talks of type d) are the hardest and – regrettably – the least rewarding. If colloquium attendants really learned some new science, they would come, at the end or even during the presentation, with very passionate, real questions. The speaker will have the unduly task to dim the eagerness as the participant very likely has gone off in a wrong direction. Rather than being happy about the newly learned stuff the participant gets irritated. He feels that his new ideas are not appreciated well enough.

The *Presentation Guide* is meant for talks of both type c) and type d). Whether your presentation has talk-c) or more talk-d) character depends heavily on the audience and on the embedding. If you give a plenary talk to a mixed audience of five hundred people, the talk must be largely of talk-c) character. If on the other end you are delivering a seminar to five experts, the talk will probably be of type d).

1.C Goal of the *Presentation Guide*

This presentation tutorial is meant for helping you to deliver talks of type c) and of type d). The audience is supposed to consist of scientists.

If you've come to read the *Presentation Guide* in order to find out how talks of type b) should be delivered, you came to the wrong place. Giving talks of type b) is far too easy.

1.C.1 Discussion groups

The author of the *Presentation Guide*, that is me, has over 30 years of experience in reading and writing scientific papers and in hearing and delivering scientific presentations. In my opinion many of the hints in this guide are crucial for delivering excellent presentations.

But it is also your guide: if you do not agree with one, some, or many of my hints, own ideas at the weblog www.sciencesurvivalblog.com. If more people agree with you, the *Presentation Guide* will be improved by implementing your advice.

http://sciencesurvivalblog.com

1.D Presentation formatter

The standard these days for presentation software is *Microsoft PowerPoint*. The *Macintosh* people will use a program like *Keynote*. The *Unix-Linux* people will very likely use *Sun*'s *StarOffice*. These applications have a lot of inbuilt facilities for making a slide show.

Another acceptable way of making a presentation with a computer is to make individual slides with a graphics program and bind these slides, for instance with *Adobe Acrobat*. The collection of slides can then be presented with a viewer program, for instance *Adobe Reader* for pdf files.

Or you can make your slide show as a collection of web pages, and use some Java slide-show software.

If a slide-showing program lacks the facility to show in full screen mode, I will strongly advice against its use. The audience will see a cluttered screen of which only a part relates to the presentation.

http://office.microsoft.com/en-us/powerpoint/default.aspx
http://www.apple.com/iwork/keynote/
http://www.sun.com/software/star/staroffice/impress.jsp
http://www.adobe.com/products/reader/

http://java.sun.com/

1.D.1 Are hints limited to users of *PowerPoint*?

A number of my hints are presented with in mind *PowerPoint* as the presenting software. This focusing is not really a limitation as *PowerPoint* is by far the market leader and because the hints are easily translated by the users of other software packages into corresponding directives.

1.E Format of the *Presentation Guide*

This tutorial is a collection of a large number of short rules, outlined and numbered in a hierarchical way. The directives often represent an independent piece of information, so that the reader can study any number of items in any desired sequence. To keep the text informal, the style chosen in the directives is sometimes as if the author is talking directly to the presenter preparing his talk. Consequently in rules one finds often "you" or "your presentation".

1.F Publication form of the *Presentation Guide*

1.F.1 Publication form

The *Presentation Guide* is available in basically two forms: as a book and as an ebook.

1.F.1.A Printed version

In the printed version the first three guides (*Presentation Guide, Email Guide,* and *Writing Guide*) are collected in one volume.

1.F.1.B Digital version

The *Presentation Guide* will also be available as ebook (a protected pdf). The pdf file will be prepared in cooperation with *FileOpen Systems.*

http://fileopen.com

1.F.2 Navigation

Navigation through the digital version is easy: there are bookmarks in the pdf version.

The printed version will be bound in such a way that it can be read hands-free.

1.G Use of the *Presentation Guide*

The *Presentation Guide* is meant to be self-contained. Studying it should be enough to produce high-quality presentations (that could still contain bogus science).

You can learn a lot from the critical evaluations of presentations given by your colleagues. If in your research group the atmosphere is sporting enough, one should criticize each other's presentation before it is shown to the outside world. Later in this guide I will give more details on how to organize these review meetings.

1.G.1 Communities

It is our plan to post on our weblog examples of excellent presentations (with comments) and examples of poor-quality presentations (with comments).

If you think you have an excellent *PowerPoint* presentation that you want to be shown as an example, contact us through our website.

If you think you have a really poor *PowerPoint* presentation and you have the guts to show it to us, as well as to other scientists, please contact us through our website. We might publish it with our comments.

Presentation quality is, to some extent, a matter of taste. You can air your opinion at our weblog www.sciencesurvivalblog.com

http://sciencesurvivalblog.com

1.H Size of the *Presentation Guide*

Suggestions for additions, corrections, or other ideas for changes and improvements are welcome. It is my intention to keep the size of the *Presentation Guide* to a minimum of about 1000 items. This size constraint ensures that scientists can read the whole text in far less than an hour. Increasing the size beyond this limit would deter too many members of the target group.

1.I Male chauvinism

In many western societies women are underrepresented in the natural sciences. This situation is highly undesirable. In the *Presentation Guide* I could have been politically correct by continuously using "he or she" and "his or her". This would make the text look ugly and I have chosen for the male solution. The reader should realize that wherever I say "he", it could well be "she" and I hope it will be 50% "she" in the near future.

1.J Training courses

In university environments courses are frequently offered on how to deliver scientific talks. These courses are a waste of time, as they are not given by active, professional researchers. Your task is to present scientific information that is crystal clear to your colleagues in the scientific community. The opinion of humanists, education specialists, professors in language, and the likes are absolutely irrelevant, and indeed invariably an obstacle.

1.J.1 Improving your English

If you want to improve your English, listen to English radio programs (like the *BBC World Service*), read high-quality intellectual magazines as the *New York Review of Books*, watch *CNN*, or watch English-spoken TV programs (without subtitles in your own language). Make sure that presentations and scientific discussions in your group are in English and not in Russian or French.

http://www.bbc.co.uk/worldservice/
http://www.nybooks.com/
http://www.cnn.com/

1.K Conventions of the *Presentation Guide*

Words that are in *italic* represent (deposited) names of brands, companies and/or computer programs. Examples: *Acrobat* and *LaTeX*. At the end of the *Presentation Guide* proper credit and (web) address information will be listed regarding these (deposited) names.

1.K.1 Double quotes

Double quotation marks indicate quotes, either from text or from speech. The text inside the quotes will be in red color (not visible in the printed version).

1.K.2 Single quotes

In this booklet I use single quotes to indicate a 'strange' word, or a regular word occurring in an unusual meaning. Instead of single quotes I could also have used the word "so-called".

1.K.3 Vague qualifiers

Language can be beautifully vague and subtle. On the one hand these ambiguities can lubricate communication. But on the other hand this lack of accuracy frequently is irritating. Take qualifiers like "often" and "sometimes". I will use them often(!). To make sentences containing such a qualifier more specific I will in many cases (95%) have this qualifier followed by a percentage between brackets. This proportion is only a rough estimate, but it will be specific enough to offer the reader an idea about the frequency of the events I am discussing.

1.K.4 Commandments

I experience around me that a large fraction (90%) of oral presentations have major pedagogic problems. These lectures set wrong examples. Such worn habits have to be eradicated from the scientific community. As a result you will find in this guide many sentences that start with "Do not do X" or "Avoid X". X refers to a bad practice that I have noticed over and over again.

1.K.5 To-Do list for you, readers of this Guide

A small number of hints refer to an 'action' on the side of the reader that in my opinion should be carried out as soon as possible. To achieve this reaction I have put an arrow (⇒) after the item, indicating to the reader that he should put this item in his To-Do list *now*.

1.L Legal disclaimer

I will regularly mention third-party commercial products that I found useful when preparing a scientific lecture or presentation. I hold no responsibility whatsoever in case you use any of those products.

1.M Slick presentations

Selling a new commercial product is entirely different from presenting a scientific colloquium. Fancy colors and fancy animations that do not serve any educational purpose are out of place in scientific talks. Slick, commercial, but hollow presentations will raise irritation in your audience. In some parts of the USA it may be appreciated that you promise to solve the energy crisis, win the war on cancer and eradicate Alzheimer's disease. In the rest of the world you will be considered a charlatan with a big mouth.

If your presentation is on the dull side but crystal clear, you have done a wonderful job. Your profession is not that of a salesman, but that of a scientist.

1.N Preparation time

You should have spent quite some time on the preparation of your presentation, but not all your time. The more you follow the hints in the *Presentation Guide*, the more you can reuse your old slides.

As a rule of thumb I would estimate that the preparation for a new, half hour talk could be completed in two full days. Preparation requires concentration. Do it at home, or close your office door, and reserve a block of time in your calendar.

2 PROCESS OF PRESENTATION

Here I discuss all aspects of your presentation not related to content. The content will be treated separately.

2.A Conferences

Conferences are usually organized by expert organizers. Ingenious time schedules have been developed by them to make the conference proceed as smoothly as possible for participants, speakers and exhibitioners.

The weak points in the organization are the chairmen and the participants. Chairmen often (50%) are not able to handle assertive speakers who go way over their time, or cannot deal with questioners monopolizing the question period. You, as a speaker – if you are well-prepared – can help your chairman, your audience and yourself. Be assertive. Later on I will present a number of hints on how to achieve this control over the situation.

2.A.1 Parallel

If you know that the speaker before you is a no-show, do not arrive too early yourself. Otherwise you risk becoming the victim of a chairman advancing the schedule. People who show up right on your scheduled time, and especially for your talk, may discover that you have already delivered your speech.

Do not leave the room immediately after you have given the talk. Do not stash away your belongings loudly while the next speaker has just started.

2.A.2 Plenary

Be very, very careful with the size of your slides. A five-hundred man audience in an ugly, large, noisy American hotel room (*Baltimore Room*) with a level floor will make the lower 50% of your slides invisible. Speak slower. Speak louder. The audience should recognize that you prepared the talk especially for this plenary occasion. Do not be stupid: preparation could not have consisted of a mere editing of your slides of one of your recent seminars.

Questions after plenary talks serve no purpose. Questions will drown in the noise of the audience. They will spoil the good impression you have made. Contrary to all my earlier and later advice, here I recommend to use as much time of the question period as possible for your talk. If you get a question, repeat it to the whole audience and answer as if you are continuing your talk.

2.A.3 Preference schedule for your talk

When you are invited to speak at an international symposium you should be proud. But optimize your performance. Being nervous for six days because you have to deliver your talk on the last day is not a very enjoyable situation. The best days for delivering your talk are the first two days.

> So when you are invited and willing to accept, or when you submit an oral contribution, tell the organizers way in advance your preference for speaking on one of the first days.

The more senior you are the more stubborn you can be with this request ("I have to teach at the end of the week" or "I have to go to another conference, already agreed on a long time ago").

2.A.4 Leaving and entering audience

Large conferences are chaotic, especially those with numerous parallel sessions. The attendees will be moving in and out of the room, even in the middle of your talk. The atmosphere will be hectic. Be prepared for large numbers of people leaving right at the moment that your talk is announced. Do not let it reflect on your mood.

The amount of people attending a scientific presentation reflects more the social status of the speaker than his scientific quality or the newness of his speech. Do your utmost for those who stay. Do this even if only the next speakers of the session are left as your listeners.

> Never make remarks at leaving or arriving people.

2.A.5 Position in auditorium

You should have inspected the room where you give your talk long before (at least an hour, preferably a day) your talk is scheduled. Try to spot the best seat. This is always an aisle seat (you can leave and come whenever you like). Reserve this seat by putting some of your inexpensive belongings on it (for instance the useless conference linen bag with all the tourist promotion material). This reservation ensures your seat will always be there for you, until you remove your belongings.

2.A.6 Hotel

If you go to a conference, it is usually organized in a large and expensive hotel. It is very convenient for you to stay at the hotel in which the conference is taking place. Only consider booking a cheaper hotel if the conference hotel is fully booked, or when the conference is not held in a hotel. If you hotel room is at the same location as the conference you can still work on your talk after you have heard other talks. You can walk into and out of your room during the day.

2.A.6.A Room mates

Out of budget considerations your group leader might be tempted to have you share a room with another group member. If you still have to work on your talk, or still have to rehearse your presentation you should get a single room. It is not uncommon to work late at night on your talk, for instance to introduce some last minute corrections. Or to rehearse aloud late at night. If you do not have the mandate to order a single room, explain your problem to your group leader. If he does not show empathy with your problem, usually out of short-sighted budget reasons, you have a problem and a bad group leader.

2.B Single event

If your talk is at an institution where you are the only speaker, be prepared that a number of facilities (like pointer and microphone) will not be present. On the other hand, unlike at a conference, you will probably have at your disposal a black board (white board) with chalk (board markers).

Your host will always be nervous, because he does not know how many people will come to your presentation. He is afraid that the number of people present will be so low that he will feel embarrassed. Tell him that you are used to extremely low numbers of people showing up at your own institute (in many cases this is the truth anyway).

2.B.1 Group meeting

Group meetings should be very informal.

Hardly ever do students get the chance to ask real questions when they are part of an audience. In the *Senior Guide* a lot of details are given about how to organize these group meetings.

In a successful group meeting the presenter does not get further than half way through his slides. At that time half of the audience should be participating in lively discussions and asking questions. Seniors should neither dominate the discussion, nor speed up the speaker. For juniors this learning process is crucial.

2.C Evaluation committees

There are many types of evaluation committees: grant proposal committees; hiring committees for new professors or other faculty members; government-instated national evaluation committees and so on.

These assessment meetings mean real money. If you go over your time, or if you have finished your presentation way before your time is up, you really messed up your own chances. Talks of this type have to be rehearsed over and over again.

2.C.1 Crucial prior knowledge

There are a number of things you have to ask (in writing) immediately after you are invited to deliver a talk at such an evaluation meeting: (i) what the language is and (ii) whether you are allowed to use a video projector.

If you are in a large country (Germany or France, for instance) in which they do not speak English as a native language, the language the committee expects, might not be the one you expect. Straighten the language question out right before you start your slide preparations. Write your slides in that language. Still be geared up for the situation that the committee requests a language change on the spot.

Evaluation committees might have a lot of interviews with other candidates or other principal investigators on one and the same day, and they might find video projectors for a ten-minute interview a delaying factor. In that case you must have your digital color slides converted to printed color transparencies. So straighten this possible exclusion of the usage of a video projector out before you prepare your talk.

2.C.2 Acquaintance

When you enter the room you should shake hands with each and every committee member, and with all other people being present. Some chairmen do not like this social activity (as it interferes with their planning) and they might try to discourage you to do so. Overrule the chairperson. Be assertive and go to the committee members, whatever distance they are separated from you. No well-educated member can refuse an extended hand.

By shaking hands you show you have initiative.

When you leave the room you must try to shake hands again. Now the chairman might get assertive and will not let you do it. Accept that.

2.C.3 Handouts

If you have any handouts, they must be in high-quality color, and they must be stapled together. Hand them out after your talk. No member will have the guts to refuse your set. You might be the only candidate doing it. It certainly means that when they later evaluate your performance, they have all the information at hand. It also shows you are not afraid that they will scrutinize your slides in your absence. Do not say silly things like "These slides are confidential".

Do not hand the set out before or during your talk. You will lose the attention of the members as they will continuously browse through the pages.

2.D Attitude

If you behave like an arrogant bastard, you will have a hard time getting the audience on your side. Be modest, and present great science.

> Treat young scientists in the same way as you treat the 'gods' of your field.

After and before your talk, try to be available for young scientists. Do not socialize continuously with important people as this behavior will deter young scientists. Continuously look around you before and after your talk. Try to break out of your circle of friends.

2.D.1 Body movements

Gesticulation is part of a person's character. A lot of these features cannot – and should not – be changed. But be aware of your shortcomings. Do not walk too much. You may fall over, and you make a nervous impression. In addition you require a lot of the – always poor – sound system when you are continuously walking.

2.D.2 Body position

Sometimes your body position and location is fixed by the organizers (fixed microphone), but much more often (95%) you can walk around freely.

You have a choice of standing left or right of the screen.

> Choose that position where you have most interaction with the audience, and where you block least of the view.

Try to look at the audience and stand with your back to the screen. If you point out something on the screen, turn to the screen only shortly. Prevent – and this is very difficult – to turn your back to the audience all the time. It is impolite, you lose contact, and it is much more difficult for the audience to follow your arguments.

2.D.3 Dressing

Medical researchers show up in an expensive suit with tie and attaché case. Physicists like to prove that they are champion in ill-dressing ("Look at me, I am the smartest guy, and I can afford to look like this"). It is all a matter of taste, culture, temperature and weather. If you really want to prove to the audience that you are a clown, be my guest. Ivy-league universities do not appreciate a speaker dressed in tennis-clothing. Having coins in your pocket making noise, does not add to the concentration of the participants.

2.E Handouts

Avoid distributing handouts before, or during, your talk. You will lose your audience as participants will browse through the handouts continuously, and they will not listen anymore. If there is no information in your talk that you want to withhold, you can leave a set of handouts at an easy spot and inform your audience at the end of your talk. Do not bring a stack of reprints. Your handouts should make clear where your papers can be downloaded.

Later on I will discuss possible printing problems with handouts.

2.F Rehearse

A good presenter has prepared his talk very well. This preparation not only includes the preparation of slides, but certainly also the rehearsal of an important talk.

2.F.1 Without audience

> After you have finished making all your slides, you should rehearse in solitude, but aloud.

You will discover that there are two complications: the presentation of each slide takes much longer than you anticipated, and in a number of cases you can't find the right verbal expressions on the spot.

You have to learn by heart a (few) expressions that you discover you find difficult to come up with improvising.

Time yourself when you practice aloud and at the pace you think you should talk at the conference. The time you will need in reality is always longer, usually because you cannot resist the temptation to dwell too long on a subject when giving the real presentation.

2.F.2 Important talk

Rehearsing for an audience is time consuming, but absolutely necessary for an important talk. Talks become incredibly better after a thorough review in an informal meeting with a few of your group members.

2.F.2.A Reviewers

Ask a number of your group members. Only people that are outspoken and who you can trust, are eligible. The maximum number is three or four. With more members it becomes an ineffective spending of time for the participants. If other group members are interested in the content of your talk, you can deliver the talk later to the whole group.

2.F.2.B Procedure

The presenter takes with him a pad and a pen (both absolutely necessary). So does each member of the audience. A chairman must be chosen out of the audience.

The talk is being given as if the speaker is at the conference. The presentation is timed by somebody in the audience. Participants note down everything they like, or do not like, about the talk (volume, speed, body movements, content, etc.).

After the delivery the chairman takes the initiative. He discusses the timing and then he asks each member to give general comments on the talk. The presenter writes all these comments down. He must do this.

After this first round of general comments, a second round starts. Here each slide is shown again and each participant comments on what he does not like, or does like about that slide.

This whole procedure is very important, but also time consuming. It easily takes an extra 1 and 1/2 hour for a nominal 1/2 hour talk.

If the quality of the talk will be of very great consequence (plenary talk, grant proposal) this review procedure should be repeated at least once.

2.F.2.C Without audience

> You should always rehearse again without an audience after you have incorporated all input from other people.

Only in this way you know how long your talk takes exactly. Some difficult sentences and transitions you should learn by heart.

3 SPOKEN TEXT

Not all spoken text should be on your slide, as your performance should not be a recital of your slides. But spectators should be able to fully follow your scientific line of arguments without having to listen to you.

3.A Allotted time

Never say: "If time will allow ...". You know how much time is left, and you know precisely when you have to stop.

Independent of the context of your presentation, stick to your allotted time. Or, even preferably, end somewhat earlier. Never ask the chairman in public, when you start, how much time you have gotten. Or during your talk, how much time there is still left. Such questions prove your poor preparation.

3.A.1 Defend your time

You should even override the chairman if he is, in order to catch up with a delayed schedule, stealing a part of your time. Observe the exact time when you start. If you see that the chairman becomes nervous way before your time has lapsed, say to the audience, looking conspicuously at your own timing device, "In the remaining x minutes I will ...". And you should know x precisely. A large count-down timer owned by you – and not by the chairman – is very convenient. A large digital kitchen timer will do.

3.B Volume, tone and pace

Do not let your voice betray you. If you have not rehearsed your talk well, you are sure going to show your insecurity with respect to certain parts of the content by unwanted hesitation, unwanted stammering, and undesired reduction of the volume of your voice. Or you will get trapped into a never-ending sentence. Especially when you explain shaky arguments, behave like an actor and present that part of your work as rock solid science.

3.B.1 Volume

You can hardly speak too loud. Many speakers (25%) speak with a voice that is too soft. Remember that all non-native speakers (maybe including yourself) have strong accents. The audience has to get used to your funny accent.

3.B.2 Pace

Your speed of talking can hardly be too slow. Many speakers (25%) go to fast.

In some countries normal speech is fast (for instance in France and Spain). Like all non-native speakers, these rapid speakers will have a strong accent. If you belong to this category, be very careful. If you do not slow down continuously, your whole talk will be wasted. Nobody will have any idea why you are so enthusiastically upset. Especially when you get excited (for instance due to critical questions from the audience) this high temperament will become an impediment.

3.B.3 Female speakers

This item is absolutely politically incorrect. Female speakers tend to have a voice that is higher in tone than that of their male colleagues. They should practice to try lowering the tone of their voice.

A more serious problem is that some female speakers (20%) start a sentence at a much lower tone than at the end. This happens in particular if the speaker is not sure about the content of this sentence. Speakers really must train themselves to maintain the same tone and volume all through to the end of a sentence. If you really want to solicit critical remarks during your presentation you should produce your insecure arguments by stammering and by reducing your volume. Rehearse. Learn some sentences by heart.

A number of tips for female speakers have been published recently in *Physics Today*.

http://www.physicstoday.org/pt/vol-58/iss-2/p54.html

3.B.4 Nervousness

It is quite normal to be nervous. You probably have to talk in a non-native language. This means the risk of stammering, speaking unclear, and of speaking too fast, even when you are quite sure about your subject. You probably know yourself well enough to realize whether you have this tendency (or you should be told during the group review of the talk).

3.B.4.A Speech support

The best help against nervousness is a thorough preparation. After that, you know much more of your subject than anybody else in the world. Learn some phrases by heart.

Speech therapists can teach you some tricks on how to speak clearly and slowly. Important is that you are aware of your shortcomings.

Regularly a scientist gets carried away during his talk. Some people ask questions, which delay him and he falls back into his old habits of talking much too fast, swallowing half his text.

3.B.4.A.1 Speech therapists

What are the tricks that speech therapists use to make you aware of your bad speech habits? This can for instance be a rubber band around the hook of your desk phone. This out-of-place object will always remind you when you pick up the phone. The rubber band says to you: "Slow down, speak loudly and clearly".

You can put a 'rubber band' on your slide. Put a small picture of a ball or the like somewhere on the bottom of each slide. Nobody in the audience will notice it, or notice its significance. But for you it is a continuous warning to slow down, to speak slowly, loudly and clearly, and to keep contact with your audience.

3.C Language

You are very likely a non-native speaker and in any case your audience will consist largely out of non-native speakers. Keep this always in mind.

3.C.1 Stopgaps

Please do not use these awful stopgaps as "actually", "basically", "at the end of the day", "gut feeling", "lo and behold", "of course", "Now, we will shift gears ...". For eradication of these habits on your side, you need the review of your talk by your group members, because for you, the usage of stopgaps has become a second nature. Nobody outside your own group will ever tell you. All what you will hear is "Great talk, John, fantastic".

3.C.2 Synonyms

Pronunciation of English by foreigners can be very problematic. You might think that you refer to "18" (eighteen) degrees Celsius, but many people in the audience will hear "80" (eighty). In all those cases that easily give rise to ambiguity, help your audience. "The temperature is eighteen – that is one, eight – degrees Celsius".

You should not be surprised that when you tell your audience that you "think the internet is a good idea", your audience understands that you want "to sink the internet".

Use synonyms for difficult to pronounce words. Say the same sentence twice, but in different wording.

Exercise: find five (three + two) homophones (different meaning but same pronunciation) in the following sentence: "Our bear cannot bear to be bare at any hour".

3.D Scientific level

The scientific level of your talk will depend heavily on the embedding of your presentation, conference, seminar, or colloquium. There is one golden rule: "A talk cannot be too simple".

3.D.1 Seminar

The audience of a seminar can be a collection of experts. Nevertheless, be careful. If there are (under) graduate students, or if the audience counts more than ten people, they cannot all be specialists in your field. In that case do not expect them to know much about your field of expertise.

Remember the golden rule "A talk cannot be too simple".

3.D.2 Colloquium

If you are invited by a colleague to give a colloquium at the department or institute where he is working, be watchful with the level of your presentation. Your colleague and his group members are experts in your very own field. That is why they invited you. Let us call the field X. Do not make the mistake of expecting the audience to consist fully out of experts in X. Do not prepare a highly specialized seminar explaining the details about X. Often this would be a big mistake.

Your audience will very likely also attract people with different expertise (Y, Z, etc.). The Y and Z people are very curious about what their local people, the X-experts, are doing. So prepare your talk for Y and Z.

3.D.3 Plenary

Plenary talks by definition have a very mixed audience. Your talk must have a general character.

In many cases (90%) the delivered plenary talks are far too difficult.

3.D.4 Popular talks

Make it simple. Then simpler. Then simpler again. In your daily life you are talking to specialized nerds. That is your world. You are functioning locally in a mutual admiration society. You have no idea how far your daily discussions are removed from the daily life of many other scientists, leave alone laymen. I know that your research field (as well as mine) is the most important in the world. But accept and expect that in plenary and popular talks many attendants do not have the faintest idea about the knowledge you consider trivial textbook stuff.

In many cases (90%) popular talks are far too difficult. Popularizing science is considered to be more and more important these days. If you do it, do it with all your heart and at an extremely low level. Do not drown them, even if they would like that.

3.E Too much information (TMI)

Listening too a talk is tiring. Listening to a talk with much information is even more tiring. There is only a very limited amount of information people can absorb and remember. So please: do not present too many conclusions. Do not show too many figures. TMI is a disaster.

3.F Funnel

A funnel is a fantastic metaphor for starting wide and ending narrow. It is never a fault to begin your talk too general (it is a fault to stay too general). Do not be afraid that some of your colleagues will get bored when you start off in a general way. Colleagues like to see how you position their field in a general context. They might even learn from the way you do it.

When people hear a talk on a subject for the first time, they will want to know how important this field is. For outsiders – this includes everybody except your ten international colleagues – it is always difficult to grasp the importance of a field. Never assume that your audience fully understands the importance of your discipline. Use 'secondary' arguments. Like: "This theory earned Philip Anderson the Nobel prize in physics". Or: "After this discovery, hundreds of scientist started to work on this subject". Or: "The holy grail in this active field is ...". Or: "The number of papers on this subject appearing in *Nature* and *Science* these days is almost countless".

3.G Tell one story

Tell only one story. If you tell two or three stories it is clear to the audience that you did not know how to fill your allotted time.

A real disaster is the speaker who, when he is almost at the end of his talk, asks the chairman "How much time do I have left?". After the chairman's answer, the speaker announces – irrespective of the chairman's answer – "This means I still have some time to speak about a different part of our work". This irritates the audience tremendously. They are already exhausted and they will have to undergo an additional set of *introduction, results, conclusions*, and *outlook* slides.

3.G.1 Operators

Telling many stories is the trademark of an operator. An operator can be the director of an institute or a large scientific facility. Or he is a group leader of a group of over thirty members. His kaleidoscopic talk contains many talks, usually with different-styled slides. Everybody in the audience realizes that the speaker has collected these slides from a number of his subordinates one or two days earlier.

3.G.2 No distraction

Goethe already knew in 1789: "In der Beschränkung zeigt sich erst der Meister" (with as next line: "Und das Gesetz nur kann uns Freiheit geben"). So limit yourself. We know you are smart, but don't prove it like I want to do here.

Don't show off with your erudition (neither in speech nor on your slides) if its significance is only secondary. The audience gets distracted, and you might get trapped in a side-tracking argument by somebody in the audience calling your bluff. Produce only those things that are relevant and only if you know you can answer questions about them.

3.H Social behavior

Natural sciences have a large social aspect to them. You have to fit in the community. You need your colleagues, and they might need you. There are a number of social rules you have to live by.

3.H.1 Credit

You will understand these rules better when you realize that all scientists, including yourself, feel grossly misunderstood. You can remedy this depressed feeling of your colleagues temporarily by publicly giving them credit. There is no harm in overdoing this at a conference. It will cost you nothing. After a day, everybody has already forgotten your talk. Real priority claims have to be in writing, in scientific papers.

3.H.1.A Opening remarks

Start with showing your gratitude towards the people that have invited you. If you are giving a seminar, say something positive about the institute where you are giving your talk. If you talk at a conference, say something positive about the organizers of the conference.

3.H.1.B Closing remarks

End your talk by showing your gratitude to the audience for having listened to your talk. Do not confuse your audience by soliciting questions immediately after this acknowledgement. You are doing what the chairman should do. The chairman is the moderator and may want to comment on your talk before the audience can ask questions. Furthermore, the audience is in confusion about whether or not they have to applaud. In principle the audience expects to applaud twice: first after your talk and second after you have answered all the questions.

3.H.1.C Conference

If your talk is embedded in a conference, you probably will have had talks presented before you in the same session. Refer to their presenters whenever possible, and learn their names – and first names – by heart. "As was very clearly explained by John in his excellent talk". And not: "As was very clearly explained by the previous speaker" A good examples is also: "After the excellent introduction by Gary I feel I can skip my introduction", which of course you do not do. Giving credit regarding the quality of previous talks will win you a lot of friends (among which the people who will be the referees of your next paper in a high-impact journal). This credit will never cost you, as you are not giving them any first-discovery credit.

3.H.1.D Colloquium

If your talk is at an institution where you are the only speaker and when you are invited by a local expert, give the group of the expert credit all the time. Even ask him advice during your talk.

This will help raise the local status of your colleague and thus of your own field. In addition the colleague will owe you one.

3.H.1.E Your ego

A typical American habit, that is quickly conquering Europe, is scientists talking about 'my' group and 'my' lab. If this is what you need to get established, then you are in bad shape.

You do not own your junior group members. Show respect to them. You totally rely on them. Motivate them. Let them feel responsible. Talk about 'our' group and 'our lab'. Or use the name of the group, like the 'system biology group'.

There is a famous story about a French scientist who had published a book and always commented about new discoveries that they were already published in "my book". In the end his colleagues, when they spoke of him, called him "My Book".

3.H.1.F Lack of credit

Never complain about lack of credit of your own work. Not about colleagues, or about referees, or about bad citation scores during your talk. It is all the sign of a real loser.

3.I Questions and interference

A question period can be quite exhaustive. Some aggressive Israeli, Dutchman or Russian is waiting at the end of your presentation to get even (we all feel misunderstood, remember). There is a lot you can do yourself to keep control over the situation.

3.I.1 Repeat the question

A golden rule for which there is no exception:

> "Repeat the question and give the answer to the whole public, and not to the questioner".

This holds for questions during your talk, as well as for questions after your talk.

I am so disappointed that questions are hardly ever (5%) repeated by the speaker. In addition the dialog between questioner and speaker develops without any involvement of the rest of the audience.

3.I.1.A Advantages

If the speaker does repeat the question, there are a number of advantages.

3.I.1.A.1 Audience participates

The important people usually sit in the front rows. When they ask their question they will never realize, or they do not care, that people in the back cannot hear their question. By repeating the question you draw the rest of the audience into the debate.

3.I.1.A.2 Gain of time

By repeating the question at length you gain time, time you need to find the right answer.

3.I.1.A.3 Reduce effect of question

By answering to the whole audience (pan your head slowly back and forth) you have taken back the initiative. You have explained his critical question and your answer to the whole audience. The man that asked the question does not play a role anymore. Do not look at him. Never engage in a two gentlemen's dialog.

If, after you have answered the question, you come back to the interpellator and ask him "Does my answer to your question satisfy you?" you really are asking for trouble.

3.I.2 Politeness

Never show your irritation. Try to create a situation in which people in the audience correct the situation.

Never point at somebody in the audience (saying "I explained this to you before", while at the same time pointing at him, is a deadly sin).

3.I.2.A Crackpots

Somebody in the audience might ask you "Don't you think that Einstein's theory of relativity is wrong, whereas my theory ..."

Answer very seriously with something like "Well, to be honest I am not aware of your important contribution. Why don't you show me at the end of my presentation the reference to the international journals where you have published this?" Spot at the same time somebody else in the audience that is eager to ask a question, and – if necessary – overrule the chairman by explicitly soliciting a question from that person: "I see you have a question. Please go ahead."

3.I.3 During talk

In some presentation formats it is permitted to ask questions during the talk. This situation is a challenge. It easily gets out of hand. You can also ask the audience a question yourself. The answers you get, or the questions you get, allow you to gauge whether you already have lost your audience.

3.I.3.A Test of audience

Do not embarrass your audience by testing their knowledge. A real social disaster is to test their level of understanding by asking a double-choice question and request them to answer it by raising their hands.

3.I.3.B Disturbing expert

What do you do if an expert question is asked, to which you know the answer, but only you and the expert are interested in the answer. And the remaining ninety-nine people will get lost if you do answer this question. "This sure is an expert, but highly technical, question. You really must be an authority on this subject. I am quite willing to discuss this after my talk. But I am afraid it would be too much asked from the rest of the audience to answer right now. So, I hope you will allow me to proceed, but again, thank you for the excellent question". Or something like that. But do not answer that question.

3.I.3.C Too many questions

If you get far too many questions during your talk, you should, by looking at the chairman, say "I am very happy with all these questions, but I really have to go on now. And I will probably use part of the question time to be able to deliver the rest of my talk."

3.I.4 At end

Even if you had questions during your talk, you will get questions at the end. If you have gone way over your time, the chairman might not allow any questions. I hope he will do that, because it would teach you the lesson to keep to your time.

Again: always repeat the question for the whole audience. Avoid having a private-like conversation with the questioner. Do not answer the question by continuously looking at the questioner. Answer the question as if it is part of your talk. So look at different parts of the audience while answering.

3.I.4.A Comments

Often (50%) questions are not asked to get an answer, but for showing that the questioner does not agree with you, or is just plain jealous. It is not really a question, but it is a comment ("I did it all before you did").

If the comment is really unpleasant or unfair, just answer with "Thank you for this comment. Are there any other questions in the audience?" If you decide to do answer his nasty comment: Answer his question without ever looking at him. Always stay very friendly and polite. You will get the audience and the chairman on your hand.

Some nasty questions you could – and should – anticipate. Have some slides ready (put them after the last slide of your talk) to back up your arguments.

3.I.5 Hostile participants

To the outside world, the community of scientists is a peaceful community, readily sharing credit and information. The truth is different. The community is just like normal life, with all the positive and negative phenomena, like jealousy, theft, lying, gossip, etc. What makes science unique is that the world 'out there' is objective and the set of rules called the 'scientific method' will in the long run separate bogus from the really good stuff. However, this in the long run, can indeed be very long. In the mean time you have to defend your 'baby' regularly.

Do not take revenge at a nasty interpellant. He is very likely a singularity. He will speak too fast. Get a red face. And use technical terms, so by default the rest of the audience is on your hand. Answer his opposition by talking to the whole audience. Repeat the question. Do not let him interrupt you when you repeat the question. Do not let him repeat his arguments. Answer decisively, clearly, loudly, and very slowly.

3.I.5.A Priority claim

You will always have people in the audience that claim that they have done it before you ("This has all been done in the former Soviet Union" or "This has all been done in the 1940's by the microwave engineers at MIT").

A good answer is the following: "I have heard this claim several times before (in this way you show he is not original). I have asked on those occasions to give me the reference where this work has been published before. Naturally, I will give due credit when this will be pointed out to me. But up to now nobody has come up with a literature reference to back up this claim. Are there still any other questions?" Now the monkey is on the back of the hostile participant.

3.I.5.B Trap

A good way to neutralize a hostile participant is to trap him. You ask him something like "Do you claim that you can prove this ...?" When he answers affirmative, you reply: "You just got yourself a new *Letter to Nature*. If I were you I would write this up immediately".

3.I.5.C Backup slides

It is always good when you expect strong opposition to have special slides available to prove your point. Put these backup slides after your last slide. If you do not need them, nobody will ever know that they were there.

3.I.5.D Humor

A good laugh is the following. You prepare a slide with a long title on the subject for which you expect the opposition. For instance a slide with as title: "This has not all been discovered before in the former Soviet Union ..." With the reasons given on the rest of the slide. You show this slide only if the anticipated opposition really materializes.

3.I.6 Hostile audience

A whole hostile audience should not come as a surprise for you. You know when you are invited in the lion's den. Rehearse very well.

An interesting trick is to power down your laptop right after your talk. Questioners cannot ask you any longer to show a particular slide.

Keep on smiling. Keep on staying friendly. Do not let them steal your talk by continuous interruption.

3.I.7 Evaluation committee

Presenting for an evaluation committee is difficult. These are once in a lifetime chances, and you can blow it all. But you also have an opportunity to gain a lot.

Committee members know a lot about you. They had to read your 20-page proposal, or your job application.

3.I.7.A Atmosphere

Be prepared for some serious opposition. The members want to show to each other that they really know what they are talking about. And it is not uncommon that they will fight over your back. Never accept an insult. Remain polite under all circumstances.

Be prepared for a serious reorganization of your talk. If you do not agree, say so, but very friendly and persistently. Talk slowly and decisively. Try to get, keep, or regain the initiative. This is the only situation where some character of talks of type a) is allowed. (See 1.B.1)

Set traps for the members. Hesitate slightly on a subject you are absolutely sure about. If they take the bite you can show off your knowledge. Give them hell.

3.I.7.B Hostile questions

Anticipate questions and comments like: "Why should one be interested in this?" "This has all been done before." "This sounds like an iteration of your earlier research." "What results will make you happy?" "Is your ambition not too high?" When you deny, they will ask "Is your ambition not too low?"

If you know your field well, you can anticipate the majority (75%) of the nasty questions. Do not be arrogant by using sentences like: "This really is a good question." If they ask a silly or wrong question, blame it on yourself: "Apparently I have not explained this point very well. Let me put it in another way."

3.I.7.C Divide et impera

If a member is persistently nasty, try to neutralize him by showing that he is attacking the field of another member of the committee. Have them fight among themselves. Try to balance your answers and comments over all the members.

3.J Requests from audience

After your talk ends, listeners might ask you for a – hard or soft – copy of your presentation. Be careful. As long as your *talk* contains unpublished data, there is no problem. But when you give away digital or hard copies, or when you post your talk on a website, you're in a whole different ball game.

3.J.1 Unpublished material

If you have slides with brand new, yet unpublished material, replace those slides with almost empty slides, containing only text like "To be published". Do the same if you publish your presentation on the net, or when the conference organizers want to do so. If you have handouts ready, you must also have replaced in them the slides containing unpublished material.

Protect your material – digital or hard copy – with a large watermark. In this way you discourage them to use your material in their talks without giving proper credit. This watermark should not be removable. Protect your pdf file with password security (*no printing, no selecting, only opening*). This will not stop die-hard hackers, but the chance is very small that a colleague will be able, and will be willing to spend enough time on it, to break your protection code.

If you have unpublished material in your presentation, you should not publish your data on the web. Not as html, not as pdf.

3.J.2 Life internet connections

Never rely on life internet connections during your talk.

I have seen this fail in many cases (70%). If you really want a life demonstration, fake it. Pretend you connect to the internet and have a script running. Immediately power off your computer at the end of your talk. Nobody will be able to check the life status of your connection.

3.K Slide by slide

Giving a talk with slides is much easier than giving one without slides. The slides are your reminder where you are in your presentation.

Do not show information on a slide that you will not discuss. If you do, the audience will realize that you are recycling an old slide and didn't take the trouble of editing it for this special purpose.

You do not read your slides line by line. Just explain the information in slightly different wording. Do not present orally crucial information that is not on your slide.

3.K.1 Slide contains plots

If a slide contains one or more (not too many, of course) plots, always guide the audience through the figure. Notwithstanding the fact that your axes are labeled with clearly visible text, explain what is plotted against what. Explain what the main conclusion is from the figure.

4 SLIDES

I repeat the golden rule: "Keep it simple, and then even simpler".

Later on I will explain that whatever you do, when you start preparing your slides, start with the master slide(s).

4.A General structure

4.A.1 File size

Try to minimize the size of your presentation.

This has nothing to do with minimizing the number of slides. It all depends on how large (in bytes) your figures and graphs are. Make small presentation-quality compressed bitmap files (like jpg) out of large bitmaps (like tiff files) or out of large vectorized figures (like ps and eps).

The size minimization will help to load your slides fast (slow loading is irritating for your public) and in fast saving during preparation. Small files will also help in prolonging the life of the battery charge on your laptop.

4.A.2 Reusability

In the field of software engineering reusability is a paradigm. Structured computer languages as C++ are designed with reusability as major design goal. The 're-user' can be yourself at a later point in time, or one of your group members. To make individual slides reusable they must be loosely coupled. That is to say they must be usable without reference to other slides.

If you buy a cupboard or bed from a furniture shop, your new purchase looks beautiful from the outside or front side. If you look, however, at the rear side, or under your new bed, you will discover a lot of not-painted, not-finished woodwork. The reason is economy: why spend effort (money) on something that no one ever sees.

The same applies to digital presentation slides. You can import oversized pictures, group, cut and paste and do ugly things with nontransparent boxes. As long as it happens outside the margins or below other drawings, no viewer will ever notice the chaos.

Yet this minimal, carpenter approach limits seriously the reusability. For another user your slide will look like one big mess. I have seen examples where people put the whole text of a slide in a title box. Even *PowerPoint* stupidly suggests doing so in its template for new slides. It works, but it is awful, amongst other things because the *Normal View* is meshed up. Be disciplined and try to do away with the carpenter approach. If your colleague group members still use this messy practice, correct them.

4.A.3 Composition

Composition is a matter of taste. My credo is, the less there is on the slide the better.

Out of mere stupid laziness, many presenters, when they discover during the preparation of their presentation that a slide gets too full, start to reduce the font sizes, to reduce the line spacing, to reduce the margins, etc. This guaranteed leads to disaster. Busy slides are chaotic, irritating, and poorly legible.

4.A.3.A Landscape

Always use landscape format. In this way there is less danger of clipping. The disadvantage is that you are used to writing text in portrait (as any book and journal has an aspect ratio similar to portrait).

4.A.3.B Separators

If you have to put several objects that are logically unconnected on one slide, do what the newspaper lay-out people do: use separators, vertical or horizontal.

These separators are thick, colored lines (use narrow, filled, borderless rectangles).

By using a thick vertical separation line in the middle of your landscape slide you can still mimic portrait slides. Beware that the two parts (left and right) should be connected in content. And do not use a smaller font size when you use this separation line.

4.A.3.C Plots

The rule of thumb is to put only one figure on a slide and have it occupy about one half to one third of the slide.

The rest could be caption and explaining text. Only in this way the important 60-years-old eminence grise in the back of the audience can read your slides and might offer you a position at a prestigious university.

4.A.3.D Alignment

A lot of the clutter can be reduced by aligning text and figures as much as possible horizontally and vertically and introducing separating lines. Of course, much better is to spread it out over more slides.

4.A.3.E Margins

Use the same margins for all the slides of the same presentation. The more consistent you are throughout your presentations, the less you have to change when you reuse an old slide.

4.A.3.E.1 Top

Do not use any margin at the top. Try to move things as much to the top as possible. Even at the cost of cosmetic or composition quality. People sitting in the last rows will love you for it.

4.A.3.E.2 Bottom

Use a large, to a very large margin at the bottom, dependent on the size of the presentation room. If the room has many rows (>15) and no inclined floor, 25% of the bottom will not be visible (clipping) for 2/3 of the audience. In a very large auditorium, this 25% can get as large as 50%. With the master slide you can force yourself not to use part of the bottom: put an attractively-colored band on the bottom of the master slide.

4.A.3.E.3 Left and right

Left and right text margins should be used. It is ugly when text starts right at the beginning of the slide and/or continues right up to the right side of it (text is almost flowing off the slide).

4.A.3.E.4 Large left margin

Use a very large left margin if otherwise the slide becomes unbalanced (everything is on the left, and the right is empty).

4.A.3.F Guides

Use the guides (a movable horizontal and a movable vertical line) for all your alignments. They are very handy. Try to reset them both to a default position after you have moved them to align something that should not follow your default. In this way you are always reminded of the default position of all text as you start designing a new slide. If you put your starting text there, it will make transitions from slide to slide smoother for your audience.

You should also use the guides to align text with figures. Try to align, horizontally and vertically, as many items on your slide as possible. It improves the symmetry and design, and makes the slide less busy as the eyes are guided by this alignment.

4.A.4 Credit

Credit is an important issue in science. You should always give credit where credit is due. It does not hurt to give more than due credit. Published papers are for eternity. Giving credit in papers should be done with great care. Presentations hardly live longer than a day. So you can give much more credit on your slides than in your papers. You probably know in advance who will be in the audience.

Giving credit by only mentioning the work of a colleague, but not having his name on your slide, is not impressive.

The praised participant knows that next time when you give your talk, when he will not be in the audience, you will not give him credit. So put the credit on your slide. If the slide is part of your introduction, say something like "Whenever I present this slide I notice that ...". This will give your colleague in the audience the impression that you give him credit in all of your talks.

If you publish your presentation (on a website) or distribute handouts, you must be more careful with undue or exaggerated credit on your slides.

4.A.4.A Full references

Do not put full references on your slide.

Nobody will need those details. Nowadays, with *Google* it is very easy to find the paper. So no initials, no pages, just the abbreviated journal name and the year. Example: "Smith and Johnson, Phys. Rev. Lett. (2005)". High-impact journals can even be abbreviated beyond their standard abbreviations. Be careful with "et al." if you know that the senior scientist belonging to "et al." is in the audience. A solution in such a case is: "Smith, ..., and Johnson, PRL (2005)"

4.A.4.B Advertising own work

Science is about discovering something and then publicizing it. And keep on claiming it. Your friendly colleagues in different parts of the world are very willing to take away your discovery. There are a number of tricks to advertise your own work in your own presentation. You should do it in a subtle way. As an understatement.

Advertise your work, but do it subtly. You can, when you explain your work on a slide, put in a small (this means you are modest) but still legible (yes, you are modest, but not crazy) reference to your own work. If the number of authors is small, put all the names there and abbreviate your own name to initials, and abbreviate the journal name to the absolute minimum. This is a real sign of class: "Belly and J.L., PNAS (2005)". The only requirement is that the reference is unique enough to find it with *Google*.

If the explanation of your new findings takes more slides, put the referral there only once, on the first or last slide of the series. If you do it more than once, people will notice the duplication, and they will assume that you do not have any other papers to show.

4.A.4.B.1 Website

At the end of your talk advertise your website on the last slide. Make sure that all your papers are available there, with high-quality pdf files (not hyperlinked to journal websites, just put the files there physically, and no *Type 3* fonts please). I assume you have bought your own relevant domain name for $25 per year (which is 5000 times cheaper than the lasers you buy every year).

4.A.4.B.2 Hyperlinks

Nowadays credit and references are often given in the form of a uri (Unified Resource Identifier). Do not fully reproduce the uri if it is ugly and long. Nobody in the audience will be able to follow that link anyway. Put something more useful there: "all information and papers can be found on our website www.mydomain.edu".

http://en.wikipedia.org/wiki/Uniform_Resource_Identifier

4.B Contrast and colors

In a disappointing number of cases (80%) the contrast in pictures and text appearing on slides of scientific talks is of poor, if not of horrible quality.

Yes, I know *PowerPoint* lets you use all kinds of colors, all kinds of grading, and a large number of other splashy visual effects.

> Whatever you do with colors, check how your slide will look in grayscales (black and white).

It should still be clearly readable and visible. This is a good check on the quality of contrast.

4.B.1 Object colors

4.B.1.A Colored background

A colored background is obtained by filling a closed shape, like a rectangle or circle, with a uniform color. On top of this background a text is often superimposed. Be sure that the contrast is good (check with grayscales). In many cases (90%) I notice the use of colors with bad contrast.

When you check the contrast, beware of the problem that video projectors vary considerably in their color profile. So only refrain from checking the contrast if you are absolutely sure that you have a safe combination of font and background. I have seen regularly (20%) speakers being surprised by the bad quality of the contrast of their own presentation ("there must be something wrong with this video projector" is a poor excuse coming out of the mouth of a speaker).

4.B.1.B Graded background

A graded background is a background, for instance the filling of a box or a circle, that changes color, horizontally, vertically, or both, or even fancier. On top of this background the text is superimposed. If the color variation in the background filling is too large, the text will be poorly visible. The reason for this is that the color of the font is either light or dark, whereas the graded background usually varies from light to dark.

There are two solutions if you do want to use a graded background: (i) either use a background that varies only in the same type of colors (from one dark color to another dark color, or from one light color to another light color). Or (ii) use the cumbersome solution, that you will find in many glossy magazines: also vary the color of the font, in an inverse way compared to the background.

When you check the contrast of a font, beware of the problem that video projectors vary considerably in their color profile. So only refrain from checking the contrast if you are absolutely sure that you have a safe combination of font and background.

4.B.1.C Background texture

A background texture has the same problem as a graded background: it has a varying contrast and will destroy the contrast with fonts and figures placed on them. In addition it is distracting. A texture should only be used if it reflects a real structure that you are trying to convey.

4.B.1.D Background picture

For a background picture the same holds as for the graded background and for the texture. But with a picture, the situation is even more dangerous. Apart from destroying the contrast, a background picture is always distracting. Never, ever use it.

4.B.2 Slide background

Every drawback of colored, graded or textured background of a shape holds also for the situation where the whole slide has this background.

There is however an additional large disadvantage. Presenters get bored with their eternal white background and change it into for instance blue (very popular these days). Even if they are very cautious with contrast, there always remains a problem. Invariably the presenters have to import figures, graphs, and/or animations. The (blue) background serves as the canvas. However, many of these imports will originate from situations where the background was white (for instance every scientific journal still prints its content as black on white). As a result all these imports have white boxes around them. And the desired result of a nice, homogeneous background is lost. It has become a hodgepodge.

Use a very dull background for your slide. This will be plain white in many cases.

4.B.2.A Acceptable background

The only acceptable way of using a background different from white on your slide is to adjust the background of all your imported material on that canvas.

That can be done with graphical programs where you can replace colors (in practice it only works if the background is homogeneous).

If you get bored with a white background, you can also invert all colors. At least that has no patchwork.

4.B.3 Dangerous colors

Professionals in the graphics industry know that faithful color presentation is a nightmare. Each device, like printer, video projector, monitor, or screen, has a different color profile. These color profiles can be imported on your computer, but using this feature requires years of graphics experience. Beautiful dark yellow on your monitor shows up as ugly light yellow on a video projector.

There are a number of dangerous colors. Yellow and orange are among them. Fluorescent colors are always dangerous. When I say dangerous, I mean to say that they mess up the contrast: fonts in these colors are hardly visible, or fonts on a background of these colors are hardly visible. Do not ever use them.

Do not experiment with colors. I have seen presentations by established scientists that were awful contrast-wise. The speakers themselves are most surprised when they look at the screen: "Here at my laptop (pointing to their laptop to prove their point) the colors come out fine".

Do not experiment with colors. Contrast means a combination of safe dark color (black, dark blue, dark red) and a safe light color (white, light grey).

4.B.3.A Default colors

An additional problem with color contrast is that lots of computer programs (like plotting software) generate figures in colors with default colors that are dangerous. Always adjust these default colors.

4.B.4 Shadowing

Special effects like shadowing, either of objects or of fonts, do not add any information and do not make your presentation more enjoyable. It distracts. Leave its use to the amateurs.

4.C Text properties

4.C.1 Font

Many fonts are designed for printing purposes and not for screen reading. On the web you will see that sans serif fonts as *Verdana* and serif fonts as *Garamond* are very popular. Sans serif fonts are supposed to be better for screens. I leave it to your taste what you use in your presentations.

http://en.wikipedia.org/wiki/Sans-serif
http://www.microsoft.com/typography/web/fonts/verdana/default.htm
http://en.wikipedia.org/wiki/Serif
http://en.wikipedia.org/wiki/Garamond

4.C.1.A Font size

It is very easy to use a font that is too small. Especially young researchers with excellent eyes use way too small font sizes on their slides.

As a rule of thumb I would say the minimum size is 30 pt (many of my readers will gasp now).

Font sizes are hardly ever too large. Do not vary the font size on a slide or between different slides. Avoid situations where on one almost empty slide 42 pt font size is used, and on another busy slide you have reduced the font size to 24 pt to have all of the text fit on the slide.

Serif fonts tend to be smaller than sans serif fonts (at the same font size). Correct for that size-difference if you combine these fonts in one text (you'd better not combine fonts in that way anyway).

4.C.1.B Bold

Should one use the bold version of a font, or not? This depends on the font type.

4.C.1.B.1 Serif

If the font belongs to the serif family (like *Times New Roman* or *Garamond*), you should use the bold version of your font all the time. If you do not vary between regular and bold, no viewer will notice that it is bold. They will just notice that your text is crisp and very clearly visible. Varying between bold and regular is patronizing anyway. Your message is: "Viewers, here you should really pay attention". But you are insulting them, they are already continuously paying attention.

4.C.1.B.2 Non-serif

For non-serif fonts (like *Arial* and *Verdana*) I cannot advise to use bold all the time, as bold versions of sans-serif fonts with large font sizes tend to be ugly (very thick). If they are ugly, you should not use the bold version, or better: you should use a serif font.

4.C.1.C Underlining

Underlining is ugly and very disturbing in a text. So do not use underlining. Use other ways of attracting special attention, and do not use too much of these extra attention attracters. Soon you will look like a boring schoolmaster.

4.C.1.D Font color

I have never in my life seen a color coding of text in presentations that was maintained consistently all through to the end. The color code starts for instance with blue text for important remarks, black for normal text and red for formulas. In the end it all gets mixed up. And if you re-use old slides, it will always be a mess. Use font colors if you like, but do it consistently. Do not build a Christmas tree, and besides, realize that many font colors have poor contrast.

4.C.1.E Collection of fonts

Many fonts are proprietary and cost money to use. Your operating system will have included a number of fonts that you can use freely.

If you still have (very) old versions of programs like *CorelDraw* or *WordPerfect*, the floppies or cd's will still have a lot of useful fonts on them. Just use them.

4.C.2 Spelling and grammar

4.C.2.A Complete sentences

Avoid complete sentences. Use telegram style.

Definite articles can often be skipped. Complete sentences will only arise if you quote from a text. Use large fonts, so never cut and paste one-to-one from a scientific paper. Retype the text.

4.C.2.B Periods

In presentations one uses telegram style. Do not make complete sentences. No periods after a line.

4.C.2.C Capitals

If possible, do not use capitals on your slide in the text.

As soon as you do this, people will expect periods after a sentence, and definite articles, and your slides will look like pages out of a novel. Capitals can be handy to differentiate for instance between different outlines: use headings that start with a capital (do not use capitals in the items).

After a colon (:) you can use lower case.
Your slide title can be in any case: lower, upper, sentence, or title case.

4.C.2.D Definite articles
Do away with definite articles as much as possible (so no "the", "a" and "an").

4.C.2.E Exclamation sign

Never use an exclamation sign on your slides.

One such a sign is already one too many. I regularly see slides where people use "!!!" or even more. This is an insult to your audience. It looks like a *Wal*Mart* advertisement. The only thing missing is a set of yellow starred balloons with "sale". I even see people use "!!!???". The latter looks more like a curse out of a comic book.

4.C.2.F Question mark
You can use a question mark. But only if it is a question. So not "These are the most important questions?". Unlike in French or Spanish, in English there is no space between the last word and a question mark.

4.C.2.G Space before unit
The length is 3 cm and not 3cm.

4.C.3 Alignment

4.C.3.A Align left
Always align text as much as possible on each slide. Use the guides *PowerPoint* supplies. Text is in most cases left aligned.

4.C.3.B Align right if text connects to right
If each line of a multiple line text refers to a different object (figure, graph, formula) on the right, right alignment makes more sense.

4.C.3.C Centered alignment
If your slide consists only out of one or two lines with a large font size, centered alignment makes more sense, as otherwise the composition looks unbalanced. Make all sentences of almost equal length.

4.C.3.D Multi-lined texts
Avoid multi-lined blocks of text. If you have them anyway, align them on the left. Never use full justification (justified alignment). An exception can be if the whole slide consists of only a few lines of very-large sized text. Centered alignment gives a better composition in that situation.

4.C.4 Title

The format of your title, like font and font size, should be prescribed in your master slide(s).

Do not format the title in your individual slides. Titles, and only the titles, should show up in *Normal View* of *PowerPoint*.

A real dumb idea, unfortunately suggested even by standard *PowerPoint* slide layouts, is to put a whole outlined list in the title. This reminds of the old days where autistic C-programmers put a whole program in one 'if-statement'.

4.C.4.A Frames

Do not put frames around the title. The problem with this is that the frames will be of different size on the different slides. You can use a filled box as a background for your title. But read the sections in this *Guide* about colors and contrast very carefully, because I have seen so many, many, many mistakes being made on the choice of colors.

If you use a filled rectangle as background for your titles it should be done at the level of the master slide and right at the beginning of the preparation of your talk. Changing a master slide half way your preparation does not lead to a consistent change of all slides.

Another acceptable style is to have titles stand out by putting a long colored thick line below them (this is not underlining as the lines are always of equal length, and are always occupying a large part of the slides. Remember: never ever use a top margin. I mean: never ever.

4.C.5 Text frames

Frames or boxes around text are ugly and irritating. They are just another irritating way of asking attention from the spectators. If you are stubborn and feel you have to attract extra attention in this way, use thick lines and do it consistently throughout all your slides. Try to center align boxed text.

Do not fill text boxes, because it will deteriorate the contrast and your slide will start to look like patchwork. If you are really stupid you will nevertheless fill the box. In that case, please do not fill it with a texture. That is a sure recipe for an unreadable text. Use a single color, or graded colors of only gray or of just one color. But again, you'd rather not do it.

4.C.5.A Good text frames

There is one example of a good text frame: the 'groupbox'.

In this case the frame has a header. You can find this in many *MS-Windows* dialogs on your computer. It is a way of grouping text together. This is how you do it. Put a frame with thick lines around the text lines. Put a borderless opaque white box over the left part of the top line of the frame (say from 10 % to 60%. Put a text box on top of this last white box. You might have to adjust the display order (bring forward, etc.).

If you use more of those frames on one slide, then left align the frames if they are above each other, or top align them if they are next to each other.

4.D Outlined text (bullets)

Outlined text or lists are a crucial part of any presentation. Try to vary their outlook. If you present slide after slide with a collection of the same bullets, it gets boring as hell. *PowerPoint* allows you to vary font, size, color, or picture, and much more of the 'bullets'.

In addition you can always lighten up your outline by displaying next to it a picture, or animated picture, or sketch relevant to the content of the outline.

4.D.1 Header

Very likely your outlined list needs a header (without a bullet). Do not assume that the slide title is the header. The title text is logically decoupled from what is on the slide. If necessary, just repeat the title as a header for your outline. The header can be followed by a colon. All the bulleted lines should be indented to the right with respect to the header.

The (numbered) list should preferably start with lower case. The header can start with a capital.

If you have two outlined lists below each other, the two headers and the indented lines should be aligned.

4.D.2 Type of bullets

PowerPoint allows you to use all kinds of 'bullets'. Any character from any font. You can also use a picture as the 'bullet'.

4.D.2.A Balls as character

The standard bullet character of *PowerPoint* is dull and its font size invariably much too small. So if you want to keep the dull bullets (•), at least increase their size substantially so as to balance them against the font size in the outlined text.

> To increase the appeal of your presentation you can change the color of the bullets, for instance by making them all red (and by using a large font size).

If you want to use balls, or any other standard bullet character, it is much more charming to use their 'picture' version.

4.D.2.B Numbered outline

You only use a numbered outline if it makes sense to number. For instance because you have to refer to them repeatedly during your talk. Conclusions, for instance, do not need to be numbered. But if you sum up reasons why an approximation can break down, individual numbering can be important. Or if you say in the heading there are four reasons, you'd better count them to four.

4.D.2.C Other bullet characters

You can use the font *Webdings*, or any other symbol font, for special symbols like diamonds, hands, and arrows.

http://www.microsoft.com/typography/web/fonts/webdings/default.htm

4.D.2.D Bullets as pictures

Rather than using characters supplied by fonts as 'bullets', it can be much more attractive to use pictures of, for instance, shining balls or diamonds.

But also pictures of numbers or characters can look much better than their font version. You can find many useful 'bullet' pictures on the net (use *Google Search Images*). With a simple graphics program you can crop the picture to the minimum. Sometimes you need to add some extra space to the left or right of the picture to get a nicely aligned set of bulleted lines.

I often use as bullet a colored picture of the 'checked' symbol ("√").

http://images.google.com/

4.D.2.E Color coding

You may mix colors of bullets if their color has an obvious meaning. For instance red for bad and green for good (you could also use smileys as bullets, to convey a feeling about the outlined items).

4.D.2.F Formatting outline lines

4.D.2.F.1 Font

Try to use the same font (size) for all outlined lines. Absolutely do not reduce font size in order to fit a longer line on one line.

4.D.2.F.2 Line spacing

If you use a large font the default line spacing can be easily too large. If you adjust it, do it for all items consistently.

4.D.2.F.3 Multiple lines in one outline item

If your outlined text sentence is too long for having it on one line (even without using definite articles and periods and colons) you must be careful. One big mistake is the *PowerPoint* default solution to have them wrap and vertically align below the bullets. This wrong aligning destroys the whole idea of an outlined list. The eye can not see anymore where the items are and where the bullets.

You should manually introduce a hard return (<Enter>) (and not a soft return <Shift-Enter>). The hard return will cause a bullet to appear on the next line.

By unchecking the outline icon on the toolbar the bullet will disappear. Write your second line here and introduce leading spaces for alignment. Now you can change the line spacing of this second line without affecting the spacing of other bulleted outline lines (because you have used a hard return).

4.D.2.F.4 Horizontal alignment

If you align the second line of one item exactly under the first line there is the possible misunderstanding that you have two bulleted lines (with one forgotten bullet).

You can left indent the second line with respect to the first line. You have to do this consistently: the first lines of all items should all be vertically aligned, and the second lines of all items should all be vertically aligned; although the result is not very appealing I would say.

If you need more than two lines for one item you probably need to redesign your outline.

4.D.2.F.5 Vertical alignment

You should have the bullets vertically aligned and the following text should also be vertically aligned. Use spaces to accomplish this (if necessary spaces of small font size for micro-adjustment).

If you use colons (":") or arrows ("⇒") on outlined lines, to indicate that the next text is a consequence, all the consequences should be vertically aligned as well.

4.D.2.F.6 Multiple-line spacing

To make it even clearer to the viewer that two lines belong to the same bullet, you should diminish the line spacing of the two lines compared to the regular line spacing of the outline items.

This individual change of line spacing is best done by making each line into a new paragraph and unchecking the outline-speed-button on the *Format* toolbar of *PowerPoint* for those lines that should not have a bullet.

If you have trouble to get the spacing correct of an outline list, you can also break down your outline into a number of lists consisting of one bullet and one item. Now you can align them in any way you like as they belong to different objects.

4.D.2.F.7 Professional examples

If you want to violate the rules I have just described about multi-lined outlined lists, you have to look at how professionals do this. Checking the websites of companies like *Microsoft* and *Adobe* will give you lots of ideas. Do not reinvent the wheel.

You will often see that outlined lists are formatted in the following way: after each 'bullet' there is one key word with larger font size, bold and colored, followed by a multi-lined text in smaller font. In this scheme the separate items of the outline are well characterized.

http://www.microsoft.com/en/us/default.aspx
http://www.adobe.com/

4.E Tables

In the *Writing Guide* I have in detail explained how to format tables. Here I just want to add that a table has a maximum number of five columns and five rows (if you use the whole of the landscape slide (I assume you are aware of clipping of the lower part).

If a table is prepared well, viewers only need to peek at the table to already get the main message.

4.E.1 Alignment

Align numbers on the decimal point. Avoid using scientific notation. Use a font with a fixed pitch (like *Courier*).

4.E.2 Absolute numbers

Absolute numbers are difficult to compare. It is easier to normalize to the largest number and state separately what the largest number is.

4.E.3 Error bars

Pay attention to numbers that represent errors (as standard deviation). Vertically align error bars on the decimal point (even if they are lumped together with the average values, like 4 ± 0.4). If you present error bars in the table it is often more didactic to present them as percentages (like $4 \pm 10\%$). Then the various error bars can be compared directly (Comparing error bars in an absolute way is hardly ever a sensible thing to do).

4.F Math

In the example guide you will find examples where one formula is generated in a number of different ways. Check for yourself what you find the best quality.

4.F.1 Math from copy/paste

I assume you have a text, like a pdf file, with the desired formula in graphical format. If you do not have the pdf file yet, you might be able to make it (with

LaTeX, for instance). Make sure it is of good enough quality. If necessary, regenerate the pdf file (change the default settings of your pdf-making program so that bitmaps are not reduced to a very low resolution, like 75 dpi) or get a high-quality pdf file from another source.

Open it in *Acrobat* or *Acrobat Reader*. Blow up the equation you need as much as possible, to almost full screen. Copy and paste it in *PowerPoint*, and reduce its size to the size you need.

4.F.1.A Contrast

The disadvantage associated with copy-paste is usually that the contrast in pasted formulas is not very good. You can improve this by making the equations bold in the source file of your pdf file (assuming you have the source file). This will not always help as a number of math symbols do not have a bold version.

A simpler way is to enhance the contrast of the pasted bitmap with a graphical program like *PhotoShop* (for instance *Filter, Sketch, Photocopy*).

http://www.adobe.com/products/photoshop/index.html

4.F.1.B Math with *Tex*

Mathematical formulas are best formatted with formatters of the *Tex*-family. No html, xml or whatever can beat this. However, it requires some effort to put *Tex* output on a *PowerPoint* slide.

Preparing your whole slide with *Tex* is not a good idea unless your presentation is just a sequence of equations (which is often a bad idea anyway, even if you are a mathematician).

There are programs that let you directly turn *LaTeX* macros into jpg figures that you can paste into your presentation. These programs generate jpg figure files for every equation you need. *TexPoint* (old versions are freeware) does it and *Scientific Word* does it.

http://texpoint.necula.org/
http://www.mackichan.com/index.html?products/sw.html~mainFrame

4.F.1.C Math with *MathType*

If you do not want to use the *Tex* family, you can use the *equation editor* of the *MS-Office* family. I strongly recommend you buy *MathType*, with is an upgrade of the *equation editor*. If your equations are simple (and contain no vectors), you should use the bold version of the equations (*Style*: *Vector-Matrix*). Unfortunately some math symbols do not have a bold version.

http://www.dessci.com/en/products/mathtype/

4.F.2 Numbering equations

Do not number your equations on your slides.

Only in extremely exceptional cases (for instance for mathematicians) it might be useful. In all other cases it just means that your equations are too small-sized, and/or you have put too many of them on one slide.

4.F.3 Size

The font size for math should be really large. On the order of 32 pt or larger. Do not reduce the size to get more math on you slide. If you really want to show a full derivation, fine. But accept you may have to use four or five slides for it. The audience might lose the overall view, but at least they can read your equations.

4.F.4 Style

In equations, use spaces around the equal sign: so not "a=4", but "a = 4".

4.F.4.A Inline math

If you use math symbols in running text, you can implement these symbols as bitmaps and superimpose them on the text (where you put some spaces in the text), but that is silly. Any position change of your text box, or other change will put the math symbols out of position. So use if possible the symbol font, or create the whole text including the math with your math program.

4.F.4.A.1 Fonts

For a number of inline math symbols you probably will need a script font (usually at a much larger font size than the size of the accompanying alphanumeric text). I always use *Brush Script*.

Do not fabricate your own math symbols out of regular characters. I have seen that people represent a mathematical prime by the quote character and the double prime by two quote characters. This is a real sign of dilettantism.

http://en.wikipedia.org/wiki/Brush_Script

4.F.4.B Font size

If you combine text with inline math (this is almost a tautology) realize that you usually have to use a substantially larger math font than the text font. You might have to reduce the automatically increased line spacing when you enlarge the math font.

4.F.4.C Coloring equations

You can make your slides and your equations into a Christmas tree with a lot of colors. It is not of much help, and usually of bad contrast. I repeat usually of bad contrast.

Judiciously coloring equations can be helpful, for instance by coloring all equations, in contrast to text (use good contrast and if possible a bold math font). Another reason for coloring is that you might want to set out a part of an equation, for instance how it differs fro a previous equation, or where you make an approximation.

4.F.4.D Boxing equations

Putting a box (frame) around an equation is usually a bad idea. It means you have too many equations and you want a few to stand out. Boxing equations always adds up to the unrest on your slide. If you box an equation, do it well and consistently. So always the same color and with a very thick line. And you must center the boxed equation. But then again: avoid it.

4.G Graphics

In this guide we make a distinction between two types of graphical units:

- Figures not being plots
- Figures being plots

4.G.1 Figures not being plots

The graphics we are referring to here are pictures, photos, logos, sketches etc. In general their information density is not so structured and not so dense as in scientific plots. They are typically made, or edited with programs like *Adobe's PhotoShop* or *Corel's PaintShop*. Figures can vary tremendously in complexity, from very simple to very complicated.

Science presentations can be very dull. Illustrations can help a lot in making them more attractive.

http://www.adobe.com/nl/products/photoshop/photoshop/
http://www.corel.com/servlet/Satellite/nl/nl/Content/1150905725000

4.G.1.A Bitmap or vector

Your group should be organized such that any plot that is part of any group publication should be available in vector format.

For insertion in your presentation you should not use the vector format as these figures easily become too large (several megabytes). The standard format for vectorized pictures for is postscript. Convert them to jpeg with the cheap *GSview* or with more expensive full-fledged graphics editors.

http://en.wikipedia.org/wiki/PostScript
http://www.ghostgum.com.au/

4.G.1.B Bitmaps with poor contrast

If you copy/paste a bitmap that was not made for presentation purposes (like a figure directly out of a digital version of a scientific journal), it might have poor (black and white) contrast. A simple way to remedy this is to enhance the contrast of the pasted bitmap with a graphical program like *PhotoShop* (for instance *Filter, Sketch, Photocopy*).

http://www.adobe.com/products/photoshop/index.html

4.G.1.C Gif transparent

A nice property of gif pictures is that their background color can be made transparent. Use this to reduce the patchwork-character of your presentation.

4.G.1.D Collection of pictures

Old versions of programs like *CorelDraw, Microsoft Word, WordPerfect,* and *PhotoPaint* contain large collections of clipart. Just copy these files from old cd's or floppies to your hard disk. You can always use them to lighten up your presentations. Copyright is never an issue as your presentations are always for a private audience.

4.G.1.E Animation

Animated figures will be treated later on.

4.G.2. Scientific plots

Scientific types of graphs have axes (X, Y, and possibly Z). Physicists, mathematicians and chemists use these a lot. They are typically made with plotting programs like *Origin* and *SigmaPlot*. They are the primary form of data presentation. In a normal physics presentation it is quite natural to have 1/4 of the total information to consist out of these type of graphs. Because they are so important we will spend quite some space on them.
In the *Writing Guide* you can find many details on how to make high-quality figures.

http://www.originlab.com/index.aspx?s=8&lm=213&pid=509
http://www.systat.com/products/SigmaPlot/

4.G.2.A Rule from the *Writing Guide*

All commercial plotting programs have wrong, unacceptable settings for line thicknesses (axes, curves and tickmarks), symbol thicknesses, font sizes, and default colors (the default colors always include the bad choices: light green, cyan, and yellow).

4.G.2.B Number of figures

> Figures should be of a large size. Let us say they should take about 1/3 or more of a landscape slide.

There is only room for two figures on one slide. If you put more of them, your slide will become too busy: figure labels will become illegible.

4.G.2.C Second X-axis and Y-axis

Always use a second X-axis on the right and a second Y-axis on the top. In this way your figure is closed. An open figure easily causes confusion. This is especially true if there are more figures.

4.G.2.D Frames around figures

Frames around figures are ugly. Try to avoid them. If your slide is somewhat busy, a frame around a figure can help to separate the items on the slide. A better solution is to give the figure its own slide.

4.G.2.E Captions and titles

Plots should have either a title and/or a (legible) caption. The presence of this explanatory text does not discharge you of the duty to explain to your audience what the figure represents.

4.G.2.F Archived plots

Preferably your group archives all its publications (not just the pdf file, but all source files) and presentations in a hierarchical group folder on a local server with read access for any group member. With VPN this folder is probably readable from all over the world.

Now you can find high-quality postscript versions of all your plots. Convert them with *GSview* or with *Photo-Paint*, etc. to jpg files. Even better is to have them converted immediately after you, or one of your group members, have created them in postscript.

4.G.2.G Cleaning up plots

If the figure you want to use has appeared in one of your progress reports, or in one of your undergraduate or graduate theses, but you do not have the source files, get the pdf file of that text. That is usually a pdf file of higher quality than the pdf you can get from the published version of your paper. That is, if you have chosen the options for the pdf-maker (like *PDFMaker* or *Acrobat Distiller*) such that the resolution of figures is not down-sampled too much.

Avoid copying figures from journals.

In the *Writing Guide* I have gone to quite some length on how to make acceptable figures. The bottom line is that people very often (90%) make them with too thin lines, too small labeling, and too small fonts. Please consult the *Writing Guide* for design principles of figures.

4.G.2.H Repair figures

Often (15%) you will have to use figures from people outside your group. If they have followed the advice laid out in the *Writing Guide*, you can use the figures right away. But it will take a while before they will all follow my advice.

In the mean time you can repair figures. You can easily hide labels and numbers on the axes by putting opaque borderless white boxes on top of them. Retype the underlying text with larger and fewer numbers and with larger labels.

If you are lucky, the figures are embedded in your source file as an (COM, in earlier days called OLE) object, and you have on your computer the application installed that can handle objects of that type. Clicking on the figure gives you the server program and you can make the lines thicker and the fonts larger.

http://en.wikipedia.org/wiki/Component_Object_Model
http://en.wikipedia.org/wiki/Object_Linking_and_Embedding

4.H Obligatory slides

There are a number of obligatory slides. Using them is not boring. On the contrary, using them is a service to the audience. Do not experiment by leaving out slides that every listener is expecting. You are not an artist or a stand-up comedian.

4.H.1 Master

> When you start a new presentation, begin with designing the master slide.

(You can even define more than one master slide, but one is enough.) The reason for this priority is that changes in the master slide usually affect new slides only, or worse, they only partially effect the old slides. Changing the master slide half way through your preparation can lead to very time consuming editing sessions, as you have to correct manually all the old slides. So you cannot change the appearance of all slides by changing the master slide after the facts.

4.H.1.A Slide titles

The place holder for slide titles should be on the master slide (in that way each title shows up in the *Normal View*. The font size can easily be 50 pt. Do not use punctuation marks as colons (:) as an ending to the title. Titles are not part of the content of the slide.

The title is preferably on one line. A second line can be added in a smaller font if you want to indicate a crucial literature reference for that slide.

Left align titles and leave some space (the size of a few space characters) between the left side of the title holder and the left side of the slide. Now when you change slides the titles will not jump from left to right.

The case of titles can be any form of case: like title case, all lowercase, and so on.

Try not to use any definite articles. So not "The model" but "Model", or use "Our model" to emphasize it is original work.

4.H.1.B *PowerPoint* templates

> Never use standard *PowerPoint* templates.

They are boring at best, and often disturbing and designed with wrong colors. The standard templates obscure texts or have poor contrast. They are designed for graphically esthetical purposes, not for conveying science.

If you always use the same master slide and the same lay out of your slides, your regular audience might get bored. So when you start the preparations for a new set of slides for a new presentation, you could change the master slide. If you do so, take care that you do not violate the reusability concept.

4.H.1.C Frames and logos

If a boy has learned a new trick that first only was accessible to professionals he cannot get enough of it.

So he will put watermarks on all slides and he will put a variety of logos on them, of his university and of his institute and of his group.

Why is this terrible? Because this makes your slide too busy already, without it even having any information on it. If you really have to show all these logos all the time you must have a large feeling of insecurity. Good wine needs no bush.

> A slide should have a title and content and absolutely no affiliation advertisement.

4.H.1.D Dates and occasion

Yes I know you have done your best. Finally you are able to put a date on the master slide and the occasion ("Conference on Volcano 2007"). All this is rubbish. It will show on each and every slide and is no information. It distracts.

4.H.1.E Progress macros

Yes I know, you are now so good that you can put the progress in the left margin of each slide. Horrible.

4.H.1.F Numbering

Do not show the numbering of your slides in the title.

If you follow my advice and reduce the content of each slide, the number of slides grows. People might get tired to see that the present slide is number 11 out of 49. Showing the progress of your talk should be done in another way (See later).

4.H.1.G Bottom margin

With the master slide you can force yourself not to use a part of the bottom of all your slides: put an attractively-colored (filled) band at the bottom of the master slide.

4.H.2 Front slide

Your front slide might be on for half an hour. You are the speaker after coffee and your laptop is connected before the coffee break. Put on a non-boring slide. This could be the contents slide. Or a beautiful (animated) picture if you want to keep the audience in expectation.

4.H.3 Title slide

On the first text slide the size of the ego of the speaker is often (50%) visible. The slide shows the title and the name of the speaker. If you are a well-established speaker it is really cool, not to display your name. Or if you really must, do it in a small font, or use only initials.

If you are a junior scientist some advertisement of your name is okay. Do not read aloud "The title of my talk is XXX and my name is John Smith". You really must suffer from a large amount of insecurity and serious lack of recognition, if you this.

If the number of names on the first slide is larger than one, use the same font and font size for all of them.

4.H.4 Contents slide

Partition your talk in logically disconnected parts and show these parts in the contents slide. The slide will very likely be an outlined list with a nice picture referring to your research.

4.H.5 Introduction slide

As a companion to your spoken introduction you will have one or a few slides introducing your field.

Like any other scientist, you would like to impress the audience with how important and relevant your field is. The temptation is great to mention a large number of scientific disciplines and subjects that profit from the advances made in your own specialized field. But limit yourself.

Only mention themes you understand yourself. Only report areas that are indeed relevant for your presentation.

If you do not follow this advice you might get stuck because a reluctant audience keeps on bugging you about these items you do not want to talk about and you do not know much about.

4.H.6 Progress

The audience would like to be regularly informed about the progress of your talk. Especially when your talk is longer that 15 minutes. There are a number of ways to report the progress to the audience.

4.H.6.A Progress slide

I prefer to regularly show progress slides. A natural break to show a progress slide is when you switch from one item on your 'contents slide' to the next item.

When you are about to start the new item, use one whole slide, almost an exact copy of your contents slide with an indication where you are in the talk.

Showing progress in this way is a moment of relief for you and for your audience. You can even recap if necessary. Color coding and/or highlighting items in the contents list shows the audience what has been reported and what will be reported next. For instance, turn the bullets of the already treated subjects into 'checked' symbols ("√").

A bad habit is to show the already treated subjects on a progress slide in an almost invisible color.

4.H.6.B Slide numbering

Slide numbering is awful. Slides are sometimes numbered like "5 out of 25", indicating that you are displaying slide number five and twenty are still to come. This is silly because if you use well-designed non-busy slides, with for instance sometimes only one line of text on them – the total number of slides will appear alarmingly large to the audience.

Remember there are many people (75%) in the audience that are used to making very busy, dense slides themselves. So they expect 20 busy slides to come.

Hard-coded numbering will also mean that you cannot reuse these numbered slides without changing the numbering. This will lead to disaster. Using a macro for generating the slide numbers will also lead to catastrophe.

If you really want to display quantitative progress, use relative instead of absolute numbers. Like "completed 25%".

4.H.6.C Headers and footers

Another ugly way of showing progress is to continuously show on each slide what the subject is, as a footer or header. I have even seen situations where on each slide the contents slide is reproduced as a miniature and highlighting shows where the speaker is now. This makes the slides look busy and distracting.

4.H.7 Coworkers

A bad habit for senior speakers is to report at the very end of their talk, who their coworkers were. Often that information gets drowned because of an impatient chairman and because of an exhausted audience. It is not a very good way of showing your gratitude, to put it mildly.

Your coworkers are very important. So are your collaborations.

> The coworker(s) slide should come directly after the contents slide, unless you have put the names of the coworkers on the title slide.

If coworkers are in your audience they will be grateful to you. Do not test their patience until your very last slide.

Spelling of names of coworkers is simple. No titles, even if they are your supervisors, and even if you are in Germany (please no "Prof. dr. habil."). Just one first name and last name. No titles. And only give affiliations if they differ from your own.

If you forget somebody who is in the audience, he will remember this for years. So pay attention. And spend time on presenting this coworkers slide.

4.H.7.A Science agencies

On the coworkers slide you can also acknowledge science supporting agencies. I think there is no need for it. In the first place it is their duty to support you. Furthermore nobody of that agency will sit in the audience, so you win no credit points by referring to them.

Putting their logo on your slide is really stupid and distracting. These institutes are nationally organized and no American will know what the *CNRS* is. The acknowledgment to the supporting agencies has already appeared in your papers. Presentations are for your colleagues, not for your managers.

4.H.8 Conclusions

Do not repeat your talk again, when you go through your conclusion slide. People are already tired when you arrive at the conclusion slide.

4.H.8.A Style

Listening to a talk is wearying.

> The conclusion slide should be a quiet slide.

A real bad idea is to present the whole talk again (with figures and all) in a miniaturized form. Do not use exclamation marks, varying font sizes, varying font styles, (too many) varying colors. Do not list your references here.

4.H.8.B Combining with other information

The conclusion slide should only contain the conclusion items. Do not put any other information, like acknowledgements, on it.

4.H.8.C One-liners

Conclusions are a set of one-liners presented as an outlined list. Do not use running text lines. One conclusion item should at most take two lines. Three or four items at maximum. If you have more lines, your ego is too big.

4.H.8.D Size

> Do not show too many conclusions.

They must fit on one landscape slide. Their font size is considerably larger than 32 pt and at least 20% of the bottom cannot be used and the top is already taken by the title.

4.H.8.E New information

Do not come with new information in your conclusion. I have seen this too often (20%).

4.H.8.F Following slides

Do not come with extra slides after you have presented the conclusions ('outlook', 'future'). People will really get tired of you. Their whole body and mind are set for the end.

4.H.8.G Afterburner

An absolute deadly sin is to give a second talk after your conclusion slide.

4.I Animation

Beginners love the animation potential of programs like *PowerPoint*. They use flying text, snowing transitions between slides, rotating figures. It is terrible. Professionals use only one form: appear. That is to say, on a mouse click it appears, or disappears, that's all. Anything else is a sin.

4.I.1 Animated text

Again, if you use animated text, the only acceptable is: appear. It can be very nice to have text appear line by line, or item by item. However, there are a number of problems with this.

It has a school master aspect to it. You hide information from your public. Some viewers in the audience would like to read faster than you explain. For instance because their mind works differently from your mind.

Another problem is really severe. You have to prepare your talk much better, because your audience does not know the next line, but you also do not know. You might think that there will be another line coming, you click, and: oops, the next slide shows. You can correct for this by adding a small, almost invisible ball at the bottom. You animate this in such a way, that you know that when the ball is there, the next click will show the next slide. It just means for you an extra click for a transition, but it prevents you from an unwanted slide transition.

Yet another problem with animated text is that when you want to back up through your slides by just pushing the page-up button or clicking the page-up mouse, you will go line by line, in reverse order through your outlined text. This is irritating for you, and even more so for your audience. This inability to pick any slide you want to show has been solved with the *Presenter view* in *PowerPoint*.

4.I.2 Animated gif

To liven up your presentation you can use animated gifs. This is a standard file format, containing a number of picture frames, which are continuously looping. They are in a way funny. On the world wide web you will find thousands and thousands of examples of them. To find the right ones can be time consuming. Do not design them yourself as this will slurp time.

There are a number of snags. You cannot tune, nor rely on the frequency of the animation. If you use too many of them, the audience gets bored. It is extremely difficult to show scrolling text this way with an acceptable speed.

4.I.3 Java applet

If you need a java-applet during your talk, be careful. First you must have all java-files (.class etc.) saved together with your presentation.

Test your presentation without an internet connection.

Only then you know you have all the necessary files on your computer. With an action button you can trigger a browser that shows the java applet.

Learn how to switch between full-screen *PowerPoint* and a browser window.

4.I.4 Video

Videos can be very nice. However, their generation can be very time-consuming. Many scientific plot software packages have facilities to simplify the generation of these videos considerably.

Some videos can be played inside the presentation. For others you need to run a media player in a separate window.

4.I.4.A *Windows Media Player*

For some videos you need the interplay of an outside program like *Windows Media Player*. Be very careful with these programs. There are a number of incompatible versions (so your video might not show well on a different laptop). I have even had situations myself where on my laptop *Windows Media Player* was showing the right video but it did not project on the video projector. These things you do not solve on the spot.

4.I.5 Extra files

If you have extra files, apart from your presentation file (ppt for *PowerPoint*) you must save them with your presentation. Use a different subdirectory for them.

4.I.5.A Too many or too slick movies

Science is about content and not about the most impressive graphics. If you show beautiful animated graphics, it is clear to the audience that you are running out of good ideas to give to your students, so you let them waste their time in make movies.

In science, one X-Y plot shows more than a whole Hollywood movie.

An occasional animated presentation of data can be didactic.

However, realize that the necessity of showing an animation is usually because your data are so complicated, or so unexpected that you do not know how to interpret them.

To hide this you let your audience get drowned in your animations

4.J Navigation through slides

Unfortunately it always was painstaking to navigate to arbitrary slides. This has now been solved in *PowerPoint 2003* and newer versions.

I will first discuss the old situation. If you have to display slide 3 again after you have displayed slide 5, there are several options.

One solution is to repeat the slide. So you make slide 6 to be a copy of slide 3. For the audience this will be very smooth.

Another possibility is to put an action box on slide 3 and an action box on slide 6. You can program these boxes such that clicking these action boxes will smoothly switch between slide 3 and slide 6. Unfortunately inserting new slides in between ruins the action.

You can change the function of the right mouse button ('context menu') in *PowerPoint*. Implement the function that will allow you to navigate quickly through your slides (assuming you have given them sensible, short titles). So the right-click option should not be the version where the mouse click implements the *Page Up* action.

4.J.1 Presenter view

In the new versions (from 2003) of *PowerPoint*, you can have two different screens: the one the audience sees and the one you see. The computer on which you show the slide presentation must have dual-monitor support turned on (you must have two monitors physically connected before this can be turned on).

If you want to practice this before you actually give your presentation (highly recommended) you could connect the monitor of a desktop computer (at home or in your office) to your laptop. Just disconnect the VGA cable (usually a tightly-screwed connector) from your desktop computer and connect it to your laptop.

4.K Transition

Some people, including me, find the instantaneous transition on a mouse click from one slide to the next blazing to the eye. You can use a transition model in *PowerPoint*. The only acceptable one is *Fade through black* and do it *'slow'*. All the other possibilities are toys that will soon irritate your public. Remember that this transition model will also slow you down when you want to back up to a few earlier slides.

> As I explained under Animation, it might help to put an almost invisible sign on the slide that tells you that the next click will cause a transition to the next slide.

Uncheck the silly option that the last slide is a black slide.

4.L Software incompatibilities

4.L.1 *PowerPoint*

Compared to all other presentation software *Microsoft PowerPoint* is an absolutely superior product. However, for a number of economical reasons, *Microsoft* is continuously changing its *Office* products. It is safe to assume that in many scientific environments one will keep on using *Office 2000* or *Office 2003* for a long time. One of these will be installed on the conference laptop.

4.L.1.A Backward compatibility

Whenever you use a new version of a software product realize that its developers will try to make the files the new version uses incompatible with the files of the previous versions. They want to force you and your colleagues to buy the new product a.s.a.p. In many cases (90% with *Microsoft*) you do not need the new features.

The new program version will always have a possibility to save in the old formats (backward or downward compatibility). The new program version will always give you the warning "you will lose important formatting if you save in this obsolete format". What they essentially say is that they will lose dollars if you keep on using the old files.

But, please keep on using the old file formats. Many more people around the world will be able to read and edit your files.

If you think you need the new features of the new version, think again and try to avoid them by finding an old way of implementing the same feature.

4.L.1.B Auto recovery

On every operating-system platform people continuously lose data. Either through a system crash, a program crash, a power failure, or by clicking the wrong button when they are prompted to save their data.

The auto-recovery option of *PowerPoint* is of no avail here. The real option *Microsoft* does not dare to implement, as it would mean that they would admit that their program or operating system is unstable. This option would be to save the data file continuously, at a customizable frequency (every two minutes or so). Not in some stupid temporary recovery file, as they do now, but in the real file.

So you must implement this safety net yourself.

> First, when you start a new presentation, save the file immediately with a good name and at the right position in your archival directory system.

This means that every backup procedure will include this file.

> The next thing is to save the file every five minutes or so by clicking the floppy icon.

Or, for the keyboard oriented: <Ctrl-S>. Do it. Do it all the time. Do it hundreds of times. Never get tired of doing it.

4.L.1.C Improved auto-recovery

In *PowerPoint 2003* (and presumably in the *2007* version) the auto-recovery feature has been improved considerably. But I still advise to use the method of saving the file yourself often.

4.L.2 Proprietary fonts

When buying your computer, or an operating system, you acquire the license to use a number of fonts. It is also possible that some fonts you are using are proprietary. That means you have to pay to use these fonts. You should never use these fonts as this hampers very much the reusability of your slides.

These proprietary fonts may get into your slides without you knowing it. For instance they might be packed into figures you imported. You notice their presence when you save your presentation. *PowerPoint* will warn you that it cannot save these fonts. You should always try to replace these fonts. Sometimes this is very difficult as *PowerPoint* does not tell you where and in which slide the fonts are used (shame on *Microsoft*).

You can find them in the following (painstaking) way.

> Open a new blank presentation. Copy each slide of the present presentation into the blank presentation. One by one. After you have copied each slide, save the new presentation.

If you do not get a warning, then you know that the proprietary fonts are not in that slide. When you get a warning, you know you've got the bad guy. Usually it is enough to copy that bad slide (select all, copy, paste) for the bad fonts to disappear.

4.L.3 Missing fonts

All computer programs that have data files associated with them face the following dilemma: should they (i) put all the data in the data file, resulting in immense files, or (ii) put less data in there and assume that the computer on which the program is run will find the missing data somewhere (from the operating system for instance). The problem with (ii) is that the computer might not have the data, or even worse, it has the data, but in a different version (this nightmare is called 'dll-hell' in the *Windows* development environment).

In *PowerPoint*, the default is such that when you save a file, no fonts are saved. *Windows* assumes that the computer will have the font, and if not, it will replace it by a similar font. This will lead to a disaster when you use font-specific coding like mathematical symbols, or non-alphabetical fonts. Suddenly your presentation will miss a number of characters. This can be very irritating.

There are two solutions to this missing-font problem

(i) do not experiment with fonts, use only those for which you know that they will be available on any computer (*Times New Roman* and *Arial*) for instance and/or (ii) use embedding of fonts (in old versions of Power-Point called: *Pack-and-Go*).

4.L.3.A Embedding of fonts

When you embed fonts (Tools, Options, Save, check *Embed fonts*) (or use the *Pack-and-Go* feature of old *PowerPoint* versions) the file is saved with all the fonts included (embedded is the technical term). This has a huge complication: the file can become very large.

4.L.3.A.1 Pack and Go (Obsolete now)

Another problem that I noticed is, that when you go through *Pack and Go*, on large files *PowerPoint* easily crashes. This has improved in the new versions. So remember the rule: save your file before you do anything involved, like for instance starting the *Pack and Go* program. Find an empty directory, or create it, where *PowerPoint* can dump its *Pack-and-Go* files. After you have run *Pack and Go*, *PowerPoint* will have saved in that directory a compressed version of your presentation and an uncompress program. Find or make another empty directory, run the uncompress program and tell it to dump the uncompressed version in that directory. There you will find the large ppt file (with a short *MS-DOS* 7-3 name) for *PowerPoint*. Notice that if you should edit this file and save it normally, all embedding is lost. So if you edit it, you have to go over the whole painstaking procedure again.

4.L.3.B Old *PowerPoint* versions

You might think, "What the heck with *Pack and Go*", I am using the 2003 (or even the 2007) version of *PowerPoint* where this has all been solved. Well, you could be in for a surprise. The advice I am going to give you here should bring a blush of shame on the cheeks of *Microsoft* people: If you have *PowerPoint 2003*, then keep an old version of *PowerPoint* on your computer as well.

The reason is that *PowerPoint 2003* has become very picky, and I mean really ridiculously picky and buggy, about font licenses. If *PowerPoint* thinks you are not allowed to use a font, a font that you might have used already for years, it will turn your own old presentations into a read-only version. *PowerPoint* makes mistakes about these fonts. You are suddenly in a situation where you cannot (re-)use your own presentations. The official advice of *Microsoft* (sic) "is to open these presentations in an old version of *PowerPoint*". I have done this and I know this way works. This is not where the misery stops, though.

4.L.3.B.1 Installation advice

(Re)Installing an old version has potential dangers. The biggest is that the old version of *PowerPoint* grabs the ppt (and other file extensions) file associations. You have to reset that manually.

Furthermore, you have to test the old version very well. You might have forgotten to install some necessary features, which are only available through the installation CD. This has caused me some sleepless nights when I was preparing my talk for the next day.

In addition, you have to install the newest service pack of *PowerPoint 2003* to be able to open the files you have saved with *PowerPoint 2000*.

4.L.3.B.2 Font-saving problems

With the *PowerPoint 2003* program I have come across font-saving problems regularly when I embed all fonts. The stupidest error message I came across regularly says "General failure saving font xxx". *Microsoft* keeps silent on how to solve this problem.

4.L.4 Printing problems with handouts

There are at least two possible problems with printing handouts.

If you use non-standard fonts, you must tell the printer driver to embed the fonts. This only works if the computer has the fonts. If it does not have them, you must have saved your ppt file with font embedding.

If your slide depends heavily on parts being opaque to other parts, the printing might not show the opaque features fully correct.

4.L.5 *PowerPoint* 2007

When finishing the new version of this guide, *Microsoft Office 2007* has been released. Many, many things have been changed.

A large number of new incompatible file formats (pptx and pptm for *PowerPoint*) have been introduced, and a totally new user interface. For the time being *MS-Office 2007* is a disaster. It will take quite some time until this immature product will have been accepted by the community. *Microsoft* will have to introduce a large number of improvements, get rid of a number of bugs and a number of add-ins have to become available. So if I were you, for the time being, I would not bother about implementing *Office 2007*.

You can use *MS-Word* and *PowerPoint 2003* and *2007* concurrently (this is not the case for *Outlook*), but it requires quite some fine tuning of registry settings.

5 TECHNICAL ASPECTS

A good presenter has prepared his talk very well. This preparation not only includes the preparation of slides and the rehearsal of the talk, but also to take a number of hardware and logistic preparations. The speaker should be well prepared for a number of surprising circumstances.

5.A At home

5.A.1 Video projector (obsolete now)

If you are not sure whether there will be a video projector, tell the organizers you will need a video projector (overhead projector is outdated).

5.A.2 Laptop

5.A.2.A Power management

Charge laptop batteries at home. Disable your screen saver (quite a number of talks have been destroyed by a screen saver popping up).

Screen savers are outdated anyway.

Laptops have power management features to save battery power when you do not need the hard drive or screen. You must rehearse at home what power management setting you need. There are two reasons for this: (i) you must be able to know how to wake up your laptop and (ii) for some laptops some power management features (like hibernating) either cause a crash or at least a long wake-up time. You do not want that during your talk.

If you can plug your laptop into the mains at the conference, then turn off all power-saving features.

5.A.2.B Laptop rebooting

Be sure that you know how to reboot your laptop computer quickly.

A number of laptops refuse to reboot unless you are following a special procedure (not just <Ctrl-Alt-Del>, for instance by manually disconnecting the battery). You can almost always switch an unwilling laptop off by holding down the power button for five seconds.

5.A.2.C Laptop resolution

> Be sure that you can change the resolution on your screen almost blindly.

Sometimes you will discover that the old video projector of the provincial French university that invited you, shows only half the image of your presentation. Changing the resolution will solve this.

5.A.2.D Wifi

If your laptop does not have wireless capabilities (Wifi) installed, get a Wifi (PCMCIA or USB) card (⇒).

5.A.2.E Wireless mouse

It looks silly when you have to walk to your laptop everytime to switch to a new slide. It gets even worse when you use animated outlined lists. This continuous walking to and from your laptop hampers your freedom. You could stumble over cables or block the view of the audience.

You need a wireless mouse. Do not use a wireless mouse with an inbuilt laser pointer, because the (red) pointer is (up to today: August 2008) always too weak.

> Buy a small wireless, optical mouse.

Bluetooth is overkill for your wireless mouse and in addition dangerous as Bluetooth interfacing is a nightmare. Be sure that you buy one that you can switch off (⇒). Otherwise the batteries will run empty even if you do not use the mouse.

> Buy extra batteries for your wireless mouse (⇒).

5.A.2.F Bluetooth

If Bluetooth is not built into your laptop, buy for your laptop a USB Bluetooth hardware component (PCMCIA or USB). In combination with your mobile phone (with Bluetooth) you can now always connect to the internet if necessary in case of an emergency.

5.A.3 Accessories

A number of small accessories, that are not part of your laptop, will be crucial for a flawless presentation.

5.A.3.A Adapters for wall plugs

Whenever and wherever you travel, you should have an adapter for the mains.

5.A.3.B Plugs and sockets

Nowadays electric plugs of charging units are usable in many hotel rooms in Europe, but not in all. American plugs (sockets) are never compatible with European sockets (plugs). So transatlantic travelers have to bring with them an adapter (available in any drug store). Also for non-transatlantic travelers a set of main adapters can be indispensable. Buy them once at an airport and keep them with your laptop.

A double europlug adapter can also be very handy.

5.A.3.C Ethernet

I assume that your laptop has an Ethernet port. Otherwise synchronizing the files on your laptop with a desktop computer is a nightmare.

You will need to bring along with you an Ethernet cable with RJ-45 connectors. Your hotel or conference site might offer an Ethernet connection, which should make you very happy.

5.A.3.D Phone and modem

When polishing your talk the night before in your hotel room, you might want to connect your laptop to the internet. Here are some hints on how to be able to do that when there is no wireless hotel service.

5.A.3.D.1 Fixed-line phone

Connecting the modem (if you have one) to a fixed phone system can be an interesting experience. In a number of European countries the phone system does not use the universal RJ11 jack plugs and RJ11 sockets. In addition hotels often have installed their own system to prevent guests from using their phones for data connection. You can try to buy a suitable phone adapter cable in a local supermarket. Keep the cable with your laptop in case you visit the country a next time.

Kits for connecting phones around the world exist. See for example

http://www.targus.com/us/default.asp

5.A.3.D.2 Mobile phone

I am sure that when you bought your cell phone, you especially picked out a model having Bluetooth. So indeed your cell phone has a Bluetooth connection (infrared is obsolete now, and cell phone proprietary cables are cumbersome). Practice connecting your cell phone and laptop as Bluetooth interfacing can be a nightmare.

If you live in Israel and visit a conference in Sweden, you can always call your provider with your cell phone in Israel. However it is much cheaper and much more convenient when you have an international internet service provider that uses local phone numbers for access.

5.A.3.E Pointer

You should at least have two pointing devices. You need a telescopic pointer (like an FM-radio antenna), available in any office supply shop (⇒). And you should always have at your disposition a laser pointer. Do not rely on the organizers to supply these tools. In many cases (50 %) there will be no suitable pointer at the conference site. Or they will supply you with a weak red laser pointer with an almost dead battery and a dead switch.

You need a bright, green laser pointer (⇒).

The problem with them is that they are illegal in a number of European countries (soccer hooligans would love to have them). There exist a number of American or Asian companies that sell them (Here is one example, or use *Google*). Be sure that the one you buy takes normal batteries.

Buy a number of extra batteries and put them in your laptop case (⇒).

Colleagues at conferences will want to borrow your beautiful green laser pointer. Even the organizers tend to confiscate them, to cover up their own shortcomings. Refuse to lend it (I lost several that way). Only bring the pointer from your hotel room to the conference for the period of your talk.

http://www.z-bolt.com/

5.A.3.F Count-down timer

You will need a count-down timer with a display that is so large that you can read it during your speech (⇒). A large kitchen digital timer is okay.

5.A.3.G Flashlight

You will be surprised to discover how dark it sometimes is in a conference room. A flashlight or USB light (only if the laptop is connected to the mains) might help.

5.A.4 Print handout

You can print handouts if you expect people will want to read them. Sorry for your ego, but this is hardly ever the case. Consider the alternative of putting your presentation on the web (as a pdf, with your watermark on each page, and password protected for printing and editing).

If you print handouts, you cannot change your talk anymore after you have printed them. Do not print them in order to use them as a guide for yourself during your talk. That does not work. There will not be enough lighting and no table to put them on. Handouts can be a good choice if you give talks to a very small, but very important audience. The handouts are their take-home message.

5.A.5 Backup

5.A.5.A For use at conference

Make a copy of your presentation files on a device that is independent of your laptop. For instance a CD-ROM, a USB memory card or a USB drive. Never use a rewritable CD: it is vulnerable and requires dedicated software. Embed all fonts (*Pack and Go* in old versions).

If you use Java applets, avi, wav files and the like, put them also on the backup device.

5.A.5.B For reuse

In the *Junior Guide* I have advised to backup all your work – that is papers, presentations, and correspondence. As part of that advice a detailed directory structure was advised, that gets backed up daily, with as last node the year: /presentations/.../2005/). Save your new presentation at the right year. Your group members should have read access to that directory.

5.A.6 Personal care

Giving a talk for a large audience is a stressful happening. Strain combined with your jetlag and inevitable delays ensure that you will not be in the best shape (that is why you should never arrive at the same day as your talk is scheduled). Be prepared for a sore throat, bad head-ache and an approaching flu, and bring a pain-reliever.

Drink a lot, a lot of water.

5.B Traveling

5.B.1 Airport

If you are like me, chances are that you still have to work on your presentation during your flight. Given the short time a charged battery will function, you will need a wall plug. Just walk around in the airport terminal. Chances are (50%) you will find a wall plug that you can use.

5.B.1.A Wireless

The simplest way these days to have an internet connection is when your hotel has a wireless system (probably WiFi) installed.

5.B.2 At conference

5.B.2.A Your own laptop

5.B.2.A.1 When to switch on laptop

This is a matter of taste and nerves. I always want to be sure far before I start my presentation that the laptop-video projector combination works. But I have seen cool-headed scientists who do all this while talking to a large audience. I have also seen it fail a number of times (10%).

I very much prefer to test the video projector before the session begins. I place my laptop on the presentation table before the beginning of the session, monopolize a wall plug and power on my computer and plug in the VGA connector of the lecture room (or the other way around; the sequence can be important) and test it. If the combination works, I unplug the VGA, and leave the rest as it is.

If there are not enough main plugs you cannot use the mains, but you should still test your laptop long before your scheduled time and then put it in some safe suspend mode (or it may do this by itself, given your power management settings).

5.B.2.A.2 Start *PowerPoint*

So I assume you have started *PowerPoint* on your computer. If your presentation uses other programs (like a browser or a media player) start them also.

5.B.2.B Troubleshooting

If the screen of the video projector does not show your presentation (but displays something like "no signal on channel X") you have a problem. Modern video projectors and laptops poll the VGA regularly and find out what combination of handshaking they need. However it can also fail. In that case you have to press a special function key combination. You should know this key combination from your experience at your home institute (often the key combination is Fn-F8).

> If nothing works, reboot your computer with the VGA plugged in.

This helps in a number of cases. The polling is only done at start-up for these systems. If this rebooting still does not work use another laptop (of the previous speaker?) and get your CD, or USB device with your presentation saved on it (with all fonts embedded and a safe version of *PowerPoint* files.

5.B.2.B.1 Be the last to test video projector

A very difficult situation to handle is a video projector that adjusts its settings depending on its input. I have seen the following disaster happening several times: presenter A tests his laptop with the video projector: result OK. Then presenter B does the same thing with again an OK-result. Now presenter A starts and there is a synchronization problem (rolling image) and/or a resolution problem. Apparently the laptop of presenter B has caused an irreversible change of settings of the video projector.

5.B.2.C Lighting

The lighting in the presentation room is often (75%) too dim. This is either because some prior speaker had low-quality slides, or because chairmen like to sit in the dark. Nowadays video projectors are very bright. So intervene when the lighting is too dark. People should be able to read the conference program and easily make notes. Too dark means your audience will fall asleep.

5.B.2.D Fixed position

It can happen that the position of the video projector is fixed with screws. If that position is not optimal you cannot change it.

If however the fixing is adjustable, check the position. If you discover that a large part (> 10%) of the top of the screen is not used, adjust the video projector, or have the video projector adjusted. People in the back will love you for this.

5.B.3 Conference laptop

Conference organizers often prefer that you copy your presentation to the computer of the conference and use that computer/laptop. There are a number of snags here.

If you agree to do this, then you should realize that they have a copy of your presentation (even if you delete it without wipe-deleting it).

If you have used uncommon fonts and have not embedded these fonts in your presentation file, these fonts will not be displayed as they should. In the worst case the conference computer shows small squares where you expect characters.

When preparing your talk, you might have used a newer version of *PowerPoint* than the one installed on the conference computer. This would be a disaster.

The conference computer has its own mouse, very likely without remote control. In addition you will not feel at ease with this new computer. So, please insist on using your own computer. Even threaten not to give your talk at all if you cannot use your own laptop.

5.B.4 Sound

The sound system at any conference or other venue, will always be poor. A professional sound system (100 k$) is far too expensive for organizers or hotels.

Do not do away with the microphone ("Can you people in the back hear me" is a ridiculous question). If you nevertheless have to do away with the microphone (because of static), ask a colleague/friend to sit in the back of the audience, to test before you start, and intervene during, if you cannot be heard.

If there is a hum in the sound, make very clear that you are uncomfortable with it. The chairman has to do something about it. If you do not continuously show your uneasiness, then everybody will pretend there is no hum and your talk is destroyed. Fixing the microphone at a different position, removing a tie or taking off your jacket, might help to get rid of the hum. Less walking might too.

5.B.5 Boards

If there are boards (black or white), clean them before you start your talk. Even if you do not use a board.

The scribbles on it will distract your audience continuously. If you are planning to use the board check the presence of white chalk or black board markers. Never use light color board markers (green or red) on white boards.

5.B.6 Flip-charts

Flip-charts are horrible. Write very large. Remove the charts of previous speakers.

5.B.7 After care

After your talk and back in your hotel room, you must do a number of things. First you will have a discovered a number of typos during your talk. Correct them. Save your file and make it read-only. Delete the link to this file from your desktop.

Recharge your laptop, disconnect the batteries from your mouse. Relax.

5.B.7.A At home

If you have changed your presentation files, put the new files on your group server.

6 TEN COMMANDMENTS

At home:
1. prepare and rehearse

At conference:
2. speak slowly, clearly and loudly
3. keep to your time

On slides:
4. be consistent all through your slides
5. use large margins (no top margin, very large bottom)
6. use large fonts (>32 pt)
7. maximize contrast
8. minimize content on each slide
9. do not attract attention (no "!!!")
10. no logos, dates, numbering etc. on master slide

7 CHECKLIST

Long before conference

1. buy Wifi card and install it
2. buy wireless mouse and install it
3. buy two pointers (telescope + laser)
4. buy USB stick and install it
5. buy count-down timer
6. buy a Bluetooth cell phone and install it
7. buy a wall plug adapter set
8. buy a fixed-phone-line adapter kit
9. install on your laptop:
   - A PowerPoint
   - B Bluetooth
   - C TexPoint
   - D MathType
   - E Acrobat
   - F Paint program
   - G GSview
   - H Plot program (like Origin)
   - I As much clipart as possible
   - J A number of extra fonts

7.A The night before you leave

1. get laptop from shelf
2. start charging battery
3. synchronize data files between laptop and desktop
4. find USB stick/disk
5. backup talk twice (embedded fonts, on group server and on USB)
6. find adapter set for the mains
7. bring wireless PCMCIA or USB card
8. bring wireless mouse (+ receiver)
9. bring two pointers
10. bring extra battery for mouse
11. bring extra battery for laser pointer
12. bring modem cable with RJ-11 connectors

13. bring phone cable adapter set
14. bring Ethernet cable
15. bring count-down timer
16. bring handouts
17. bring aspirin or other pain reliever

8 ABOUT

8.A Trademarks

Acrobat, Adobe Reader, Distiller, InDesign and PDFMaker are registered trademarks of *Adobe Systems Inc.*
  http://www.adobe.com/
GSview is a product of *Ghostgum Software Pty Ltd.*
  http://www.ghostgum.com.au/
Google is a registered trademark of *Google Inc.*
  http://www.google.com/corporate/index.html
Java and StarOffice are trademarks of *Sun Microsystems, Inc.*
  http://www.sun.com/
Keynote and Macintosh are registered trademarks of *Apple Computer Inc.*
  http://www.apple.com/
MathType is a product of *Design Science.*
  http://www.dessci.com/en/
PowerPoint and Windows are registered trademarks of *Microsoft Corporation.*
  http://www.microsoft.com/en/us/default.aspx
Origin is a product of *OriginLab Corporation.*
  http://www.originlab.com/
Scientific Word is a product *MacKichan Software Inc.*
  http://www.mackichan.com/
SigmaPlot is a product of *Systat Software, Inc.*
  http://www.systat.com/products/SigmaPlot/

EMAIL GUIDE FOR SCIENTISTS

1 INTRODUCTION

The text you are reading at this moment, the *Email Guide for Scientists* (or short: *Email Guide*) lays down a set of rules that will be helpful when writing and reading emails.

The *Email Guide* is part of the *Survival Guide for Scientists* (from now on the *Survival Guide*). Occasionally there might be general referrals in the *Email Guide* to other parts of the *Survival Guide.* However, the *Email Guide* is self-contained. For completeness we present here the names of all the guides that together constitute the *Survival Guide*:

- *Presentation Guide for Scientists,* or short: *Presentation Guide*; as an addendum to the *Presentation Guide* we have published the *Example Guide*
- *Email Guide for Scientists,* or short: *Email Guide*
- *Writing Guide for Scientists,* or short: *Writing Guide*
- *Survival Guide for Junior Scientists,* or short: *Junior Guide*
- *Survival Guide for Senior Scientists,* or short: *Senior Guide*

1.A Goal of the *Email Guide*

The idea of this tutorial is to instruct readers how to (i) spend less time on their email communication, (ii) send fewer emails, (iii) receive less of them. And (iv) how to make the sent emails more concise and more informative.

The *Email Guide* is meant to make your email communication as efficient as possible. The major goal when writing emails is to send crystal-clear email messages and simultaneously minimize the time spent on those email messages. If you write your emails with care, the receiver will be very pleased. Just by merely glancing at the email sent by you he should already know what it is all about.

1.A.1. Discussion groups

The author of the *Email Guide*, that is me, has over 25 years of experience in reading and writing email messages. In my opinion many of the hints in this guide are quite helpful for scientists.

But it is also your guide: if you do not agree with one, some, or many of my hints, post your own ideas at the weblog www.sciencesurvivalblog.com

If more people agree with you, the *Email Guide* will be improved by implementing your advice.

http://sciencesurvivalblog.com

1.B Target group

The target group I had originally in mind was students – undergraduate and graduate – and postdocs in the exact sciences. From experience I discovered that more senior scientists can also benefit from studying the set of instructions laid out in this *Email Guide*.

The text will be useful for workers in many other disciplines too. Researchers in various fields can easily skip parts they consider too closely related to science, or which they deem irrelevant for other reasons.

If your social skills are far better than those of the average scientist, you might even try to convince secretaries to adopt many of the rules outlined in this guide.

1.B.1. Technical text

Some text contains some technical information that is probably only useful for those readers that are more technically oriented. This text is to be found in the last two chapters: *Security* and *Internet Protocols*.

1.B.2. Male chauvinism

In many western societies women are underrepresented in the natural sciences. This unbalance is an undesirable situation. In the tutorial I could have been politically correct by continuously using "he/she" and "his/her". As this 'bisexual' language would make the text look ugly, I prefer to be politically incorrect. The reader should realize that wherever I say "he", it could well be "she".

1.C Format of the *Email Guide*

This tutorial is organized as a large collection of short rules, outlined and numbered in a hierarchical way. The directives often represent an independent piece of information, so that the reader can study any number of items in any desired sequence.

1.C.1. Size of the *Email Guide*

Suggestions for additions, corrections or other ideas for changes and improvements are welcome if they do not make the text much longer.

The preferable way is to post a comment at the weblog www.sciencesurvivalblog.com. It is my intention to keep the size of the *Email Guide* to a minimum of about 300 items. This size constraint ensures that scientists can read the whole text in much less than half an hour. Increasing the size beyond this limit would deter too many members of the target group.

1.C.2. Publication form

The *Email Guide* is available in basically two forms: as a printed book and as an ebook.

1.C.2.A Printed version

In the printed version the first three guides (*Presentation Guide, Email Guide,* and *Writing Guide*) are collected in one volume.

1.C.2.B Digital version

The *Email Guide* will also be available as ebook (a protected pdf). The pdf file will be prepared in cooperation with *FileOpen Systems.*

http://fileopen.com

1.D How to use the *Email Guide*

The *Email Guide* is self-contained. I assume that the reader is familiar with the use of one of the popular email clients such as *Outlook, Outlook Express, Opera, Pegasus* or *Thunderbird.*

1.D.1. Navigation

Navigation through the digital version is easy: there are bookmarks in the pdf version.

The printed version will be bound in such a way that it can be read hands-free.

1.E Conventions

Words that are in *italic* represent (deposited) names of brands, companies and/or computer program names. Examples: *Acrobat, Google.* At the end of the *Email Guide* proper credit and web-address information will be listed regarding these (deposited) names.

1.E.1. Double quotes

Double quotation marks indicate quotes, either from text or from speech. Quoted text is printed in "red" (in the digital version).

1.E.2. Single quotes

In this booklet I use single quotes to indicate a 'strange' word, or a regular word occurring in an unusual meaning. Instead of single quotes I could also have used the word "so-called".

1.E.3. Hyperlinks

Hyperlinks that can be clicked are underlined and have a color different from their embedding. In the printed version hyperlinks are of no use, but are still indicated.

1.F Legal disclaimer

I will regularly mention third-party commercial software products that I use or that I find useful (or find not useful). I hold no responsibility whatsoever in case you use any of those products.

2 GENERAL PRINCIPLES

Scientists send and receive tens and tens of emails a day. If these email messages are not formatted according to strict rules, email management becomes a major part of the daily work of a researcher. But researchers should be in the lab, or should be performing a calculation, or should be writing a paper.

2.A Face to face or by email

For a number of human interactions the best way of communication is to go to each other's office, or to go to the cantina, or to get in touch by phone. However, oral transfer of information is often confusing and marred with misunderstandings. Different parties remember different interpretations of the conversations; in all cases these – unintentional or deliberate – variations of interpretation are to your disadvantage.

My advice is to settle many of your exchanges of information by email.

Email is in writing. It can immediately be put on your To-Do list. It can be communicated to other people quickly, and your email will be archived by both sides.

2.B Confirm by email

If you suspect that the agreement you just settled orally with your manager, dean, or head of department, will not be upheld in the future, confirm the agreement by email. They will hate you for it, but they will have to respect the agreement.

If your relations have indeed cooled below freezing point, you might even have to confirm by fax.

2.C Educate the world

2.C.1. Colleagues

If you receive an unclear, cluttered email from somebody who, you expect, will send you more email messages in the future, point out to him that you did not like his email. You could refer him to this *Email Guide*, possibly with explicit reference to the numbers of the items they violated.

If he keeps on sending those inefficient emails, reply stating that you will delay answering his email, or that you will refrain from answering his emails at all.

2.C.2. Educate secretaries

When I see how often emails sent by managers or their secretaries are violating my basic rules, I get depressed. It could be done so much more efficiently.

You must be very careful when telling secretaries how to do their work. They see scientists as Sunday's children: always traveling to exotic destinations, always arriving late at the lab and always making much more money than they do.

2.C.3. Educating managers

When you receive amateurish emails from managers, you should be ruthless. I regularly receive, for instance, attached pdf files that contain scanned images of typed letters of the dean's office. Really. (Text sent as an image cannot be text searched or – partly – selected.) Apparently the manager has not been able to have his organization come up with a 'home style file' (extension "dot" in case of *Microsoft Word*).

In your eternal fight with bureaucracy it is pertinent to point out to your managers what they do wrong, including what is wrong with their emails (not to their secretaries: they suffer enough already having to work for such a manager).

Do not be a pushover. Do not tolerate being flooded by time-consuming emails from the management.

2.C.4. Educating your students

When undergraduate or graduate students make mistakes in their email communication, you can never be clear enough. Tell them immediately what you do not like about their emails (their eternal spelling errors, for instance). Learning to write impeccable, efficient emails will be very good for their career. And it will help them to put more order into their other tasks as well.

2.D Internet philosophy

The rules of the internet, protocols, are voluntary standards, set out in *Requests for Comments*.

A problem with the internet and world-wide-web is that many (new) users do not know much about the philosophy of the internet. In addition, commercial software developers do their utmost to destroy the open character of the internet.

http://www.rfc-editor.org/

2.D.1. *Requests for Comments*

All the *Requests for Comments* (RFCs) can be found on the internet. The basic idea about email, and as a matter of fact about all internet data transfer, is to communicate using plain ASCII. ASCII text is mean and lean, and universal.

Commercial software developers try to destroy these standards and replace them by their own proprietary standards.

http://www.rfc-editor.org/
http://en.wikipedia.org/wiki/ASCII

2.D.2. ASCII as standard

Email communication, like all internet protocols, involves plain simple files, which are sent around the internet in ASCII, either in a directly readable form, or in an encrypted ASCII form, Base64 (so as to be able to send binary files). The basic ASCII characters are:

“ !”#$%&’()*+,-./0123456789
:;<=>?@ABCDEFGHIJKLMNO
PQRSTUVWXYZ[\]^_`abcdefg
hijklmnopqrstuvwxyz{|}~”

which includes the space character “ ”.

http://en.wikipedia.org/wiki/ASCII
http://en.wikipedia.org/wiki/Base64

2.E Email ‘programs’

The word ‘email program’ is confusing. It is better to make a distinction between an ‘email client’ and an ‘email server’.

2.E.1. Server or client

Unfortunately in the email world the distinction between server and client is shaky. One moment, the computer (better: a program running on a computer) of your employer or provider is working as a server, for instance when it receives, through your *Outlook* application, your request to send your email to somebody in the world. The next moment, when the same computer relays your email message to other computers in the world, it has become a client of those distant computers.

We will make the following practical distinction: programs run by your employer or provider will be called servers. The program run locally on your computer (like *Outlook, Outlook Express, Opera, Pegasus or Thunderbird*) will be called an email client.

2.E.2. Email servers

Your employer or your provider has at least two server programs continuously running for you (and your colleagues). One of these servers receives the email from all over the world sent to you and delivers it to your mailbox: usually this server is using one of the following two protocols: POP3 or IMAP4.

The other program (server) run and maintained by your employer or your provider, is the program that sends emails from your organization to computers all over the world. Usually this is an SMTP (or ESMTP) server.

So POP3 is for receiving email messages and SMTP is for sending email messages. *Microsoft's Exchange Server* is providing both server functionalities, albeit of course implemented in a proprietary manner.

http://en.wikipedia.org/wiki/Post_Office_Protocol
http://en.wikipedia.org/wiki/IMAP
http://en.wikipedia.org/wiki/Simple_Mail_Transfer_Protocol
http://en.wikipedia.org/wiki/ESMTP

2.E.3. Email clients

An email client, that is what is usually referred to as an 'email program', has to do two basic tasks for you:

read: the client reads email messages that are in your mailbox at the server and it allows you to transfer the messages to another file location.

send: the client transfers emails that you want to send, to the server of your provider.

The technical term for email client is MUA (Mail User Agent).

There are numerous email clients (like *Opera, Thunderbird, Pegasus, Outlook* and *Outlook Express*). They vary from having very sophisticated GUI (Graphical User Interface) interfaces to very down-to-earth *Perl* scripts, where you send email from the command line.

I do not advertise a preference for any email client, but it should allow you to store your received and sent email messages in hierarchical folders. Only in this way your inbox, and the folders related to it, can be turned into your To-Do lists.

http://en.wikipedia.org/wiki/GUI
http://en.wikipedia.org/wiki/Perl

2.E.3.A Webmail

The ever increasing popularity of the internet has caused the appearance of many forms of web interfaces to deal with your email.

A problem with any web interface is that you have to be on-line to be able to use it. There is no way of preparing your email messages off-line and then send them in one bunch when you work through a web interface.

Another serious problem with any web interface to the email server of your provider is that it will be very problematic to save your messages to local or network drives. Unless your provider or employer allows you unlimited disk space on their server, the web interface does not allow you to keep your whole email archive on their server.

Nowadays more and more commercial companies, like *Google*, allow for almost unlimited storage. However, searching through that is still cumbersome and slow.

If you can store your emails on the server and also download them regularly to your own computers, you run the risk of losing emails and of getting duplicated emails.

2.E.3.A.1 Clumsy web interface

The availability of a web interface does not imply that every browser can make use of the interface. As a matter of fact, in the battle over web browsers, web interfaces are generally made to work best only for one web browser (usually *Microsoft Internet Explorer*).

Web interfaces as a rule are unwieldy and slow. Any mouse click results in downloading a new page. In all honesty, it must be said that targeting the web interface to one browser only has the advantage that the interface can become considerably more user friendly.

2.E.3.A.2 Webmail as a last resort

Most convenient for dealing with your email are those servers that allow you to approach them in the normal way (POP3, IMAP4, SMTP, *Exchange Server*) and in addition offer a web interface. In such a situation you only use the web interface when you are abroad without your laptop or at any other place where you cannot access one of your own computers.

2.E.3.A.3 Copy of recent emails

Some email servers (for instance *MS Exchange* and IMAP4 servers) can be instructed through email clients to keep a copy of recently received and sent emails on the server. If, due to circumstances, you have to use the web interface you still have your recent emails available as a help for sending your new emails.

2.E.4. *Windows* or *Unix*

Email communication is done with and by computers. I do not want to start a discussion here on the virtues and vices of different computer operating systems.

In the present guide I will often use examples based on using *Microsoft Outlook*. This is not only done because this email client is the most popular email client. It also happens to be the program I usually use in my daily work.

This focusing on *Outlook* does not limit this guide, as the vast majority of my comments are valid for many other email clients as well. And for those hints that seem to apply to *Outlook* only, readers will nevertheless be able to transfer the implementation of my advice to their favorite email program, like *Opera, Pegasus,* and *Thunderbird.*

http://office.microsoft.com/en-us/outlook/FX100487751033.aspx
http://www.opera.com/
http://www.pmail.com/
http://www.mozilla.com/en-US/thunderbird/

2.F To-Do list

With a program like *Outlook* you can put anything (files, folders, notes) directly in the inboxes. This is very convenient as your inboxes can become a real, flexible To-Do list. Do not use any special program dedicated for managing To-Do lists. Your email inboxes will do the job much better.

If your inbox(es) forms your To-Do list – a practice I strongly recommend – a problem may arise if you cannot directly put messages in your inbox. As an emergency exit you can send email to yourself so as to be able to put a To-Do item in your inbox. (You have to be on-line unfortunately.)

2.G Raw email address

The RFCs dealing with email addresses accept many address formats. The crudest form of an email address must contain a username and domain name, like "J.Doe@microsoft.com". The username is case sensitive and the domain not. Many software developers do not seem to know of this case convention.

Your email client can decide to send all kinds of aliases along with the raw email address. The internet RFCs about email allow many different forms of embedding aliases (like first name – last name combinations) as an additional comment in the email address.

Email clients might not show you what they actually send (like *Outlook*), sometimes you may customize this.

2.G.1. Hiding email address

Unfortunately many email clients, like *Outlook*, hide the raw email addresses from the user. Sometimes you can customize the email client to display the raw address, and sometimes you cannot (as is the case with *Outlook*, unless you are prepared to delete entries in your *Contacts* folder). Being able to see the raw email addresses can give you valuable additional information, about spam, security, hoaxes, etc.

2.H Your own email address

If you have some freedom in choosing the name part of your own email address, choose the following case-sensitive form: "J.P.C.Morgan@unc.edu" (the part after the "@"-sign is not case sensitive), or "A.de.Vries@mycompany.com". That is to say:

> Put all your initials, insertions and your full last name in the name part, separated by dots.

All recipients now know all your initials exactly. People that know your initials can successfully guess your email address. In addition, you have minimized name collisions with other people at your institute.

You will never ever have to change the name part of your address anymore. If you have more providers, then try to keep the name part identical.

In a supplementary alias you should put your first name in full, like "John Morgan". An acceptable combination for SMTP servers is: "John Morgan" <J.P.C.Morgan@unc.edu>. Any recipient of this full email address has all there is to know about your name right at hand. This can be useful for them if they have to write you a formal letter, or have to put your name in a database, or if they have to look it up in one.

3 RECEIVING EMAILS

You will receive many email messages every day. Some emails require your immediate attention. Other messages are just informative and sent cc (carbon copy) or bcc (blind carbon copy) to you. You should customize your email client in such a way that you can sift through them in no time.

3.A Filtering

You can create a number of subfolders in your (local) inbox. By using adjustable filters, supplied by your email client, you can sort all incoming emails. A disadvantage associated with elaborate automatic filtering of incoming mail is that you have to check multiple inboxes to verify that you have new mail.

There are a number of repetitive email events you may want to filter out. If your group members email the whole group about their absence and presence in the next few days, ask them to put in the *Subject*-header a unique identifier. You can filter these mails immediately, using this identifier. If you are looking for some group member and you cannot find him, only then do you need to check the "Presence" inbox folder (or the online group calendar), to see if there is an email by this person, explaining why he is not in today.

Other emails you might want to filter are the ones of the type 'email to all'.

3.B Spam

Many educational institutions use standard open-source spam filters. The way these filters work is that they read all your email and they gauge their spam character by a number of criteria (an email containing

"Viagra", several exclamation marks and the word "cheap", will certainly be flagged as spam).

The spam rating will be put somewhere in the mail header (remember emails are in ASCII, so adding text is very simple). For instance a number of "*", or a number of "s" is added, where the number is a measure for the degree of spam of the message.

Next you use the adjustable filters any decent email client offers. Tell the filter that email with a header containing at least "****" is to be classified as spam and have it dumped into some innocent folder. By allowing less or more "*" to pass, you can relax or tighten your filter. A good filter will only occasionally (one per day) let through spam and will hardly ever (in my case once per half year) erroneously classify a bona-fide email as spam.

3.B.1. Commercial anti-spam software

If you have multiple email addresses (that is POP3 mailboxes) of which some are not protected by spam detecting software, you have to buy a commercial spam-fighting program. I use *SPAMFighter* to great satisfaction.

http://www.spamfighter.com/

3.B.2. Included hyperlinks

A well-known trick of spammers is to send email in html format with in it embedded hyperlinks to internet resources (images for instance). Many (older versions of) email clients follow these links automatically, that is before showing the email to you with their viewer program.

In this way the spammer site gets an automatic confirmation that your email address is valid and alive. So they will send you more spam. More and more email clients have abandoned this automatic sending of http requests associated with embedded hyperlinks.

If your email client has the facility to block downloading hyperlinks, use it. If your email client does not possess this facility, get another email client.

3.B.3. Non-commercial spam

A whole different kind of ball game is the large load of non-commercial spam. There are a number of not-for-profit organizations, like ministries, grant organizations, and governmental agencies, which, after having discovered email as a communication channel, keep on sending you emails about their activities. Their managers cannot imagine you would not want to receive their emails. They get your email address for instance from the grant organizations.

In many cases there is no way to unsubscribe to their infantile newsletters. So what can you do? You could filter your mail on the email address of the sender and delete them automatically. However, for fun you can do something more aggressive.

3.B.3.A Get even with offenders

Many of those government relations are blind and deaf to your polite request to not send you any email messages any longer. For this I have an aggressive solution that always works.

Use *Google* to get a list of about ten to twenty valid, randomly selected email addresses within the same domain name. Then send the director of the organization your complaint about him

("info@domain_of_offender" is a good guess if you cannot find the director's email address) and cc this list of those randomly selected members of the organization. Sending of such an email is always a success. The manager will be exposed to his whole organization.

3.B.4. Spam reporting

Any unsolicited email sent to a large number of people is by definition spam. If it either has an 'opt-out' or if it is a single-event email (so no newsletter), and if the addressees form a well-defined target group, the mailing is to some extent acceptable. If there is no 'opt-out', it is spam all the way.

To be able to report a spam-offender, you will need the complete message header of the offending email. Many email clients, for no reason, want to hide this header from you by all means. There are a number of ways around this limitation for all email clients, including tips on how to get at the message header for *Outlook*.

http://www.spamcop.net/fom-serve/cache/19.html
http://www.outlook-tips.net/howto/view_source.htm

3.C Viewing

The standard user interface of an email client is a GUI (child) window that shows a multi-line, multi-column view (string grid view, without grid lines) with on each line information regarding a single email message. (With *Outlook* you really have to do your best to get it that way, but it is possible: start *Outlook* from the command line with the *cleanviews* option, as "outlook /cleanviews".)

By clicking on a column header (like *From*, *To*, *Size*, or *Date*), you can sort all shown email messages according to the entries of that column.

http://www.outlook-tips.net/howto/grouping.htm

3.C.1. Adding columns

You can remove/add columns and adjust their size.

I strongly advise to instruct your email client to show a column that exhibits the contents of the *To*-message header,

If it is not displayed by default already. Even making the contents of the *Cc*-header visible by adding a *Cc*-column is valuable.

When reading received email messages, it helps a lot to know immediately whether your email address (or name in case of an alias) is either in the *To-header* or in the *(B)Cc*-message-header of the received message. If your email address is in the *To*-header and nobody else's email address is there you know instantaneously that the message is really meant for you, and not for instance for your whole institute (with an entry in the *To*-header like "all@your_institute.edu".

3.C.2. Font

Your email client regularly allows you to choose in which font information (column headers as well as the received message itself) will be presented to you. There is no way for you to know what font the sender wants you to use if the message was sent in plain ASCII.

For a number of reasons a monospace font is to be preferred. Such a font allows for consistent presentation of vertically aligned text (like columns of numbers). As you cannot know whether the sender has used such a font, there is not much to gain, to do it that way. In addition, email clients tend to be very sloppy in dealing with hard returns, so destroying carefully prepared vertically-aligned text. It is basically a matter of taste what font you use.

http://en.wikipedia.org/wiki/Typeface

3.C.3. Format

If your client allows you to choose the format in which it will show emails to you, always choose for the lowest possible format: plain ASCII or raw html. With this low-level formatting you easily detect spam, hoaxes, fraud, etc.

3.C.4. Mail check frequency

Setting your email client to periodically fetch your email is a bad habit. It conflicts with your daily routine and gives other people the opportunity to get your attention continually. Twice per day is sufficient: in the morning and in the late afternoon.

3.D Acting on received emails

You only have to react to emails that have survived all the spam filtering and the personal filtering you have added yourself.

3.D.1. No action required

Emails that you receive are either informative or they require action on your side. After you have diagonally read an informative email you can just store it in your archival system.

3.D.2. Single action required

Writing and sending an email can be done in half a minute. After that the monkey is on the back of the recipient: "John, can you send me a concept for our new proposal asap". This half-minute action on the side of the sender might require a couple of days work for the receiver.

You must discourage people sending you these types of emails.

The Golden Rule is: "Always stall your actions".

Any quick response on your side will trigger another response, and so on. If you stall, you might be lucky and the sender might already have found another solution.

If the requested action is reasonable, leave it in your inbox or file it with your other tasks. In that way it will show up in your normal To-Do list.

3.D.2.A No more Mr. Nice Guy

Refuse to acknowledge having received an email. Even if requested by the sender. Or, if you really have to do it, do it very, very late. If you are lucky, somebody else has already solved the problem. Remember: "Being a nice guy will result in an ever increasing work load".

3.D.2.B Action is writing email(s)

If part or all of the expected action is writing emails, turn to the section on writing emails in this guide.

3.D.2.C Deadlines

3.D.2.C.1 No deadline

If the email you receive contains a requested action without a deadline you might have a problem. The best thing to do is to file it on your To-Do list in a subfolder called something like "Wait and see". Do not work on it until the sender starts to complain about not having received your reaction yet. Remember: "Always stall your actions". Explain that you did not realize that the deadline was so early. Make clear to the sender what, according to you, is a reasonable deadline. After having received a clear deadline, you can reclassify it on your To-Do list with the known deadline.

3.D.2.C.2 Unreasonable deadlines

Many emails that require action on your side will have a deadline associated with it. The most important thing is that you determine yourself if these deadlines are reasonable. If the requested deadline is unreasonable, then you have got two options: (i) you immediately reply that you will not accept the deadline and that you will not carry out the requested task. The problem is now back where it belongs. Or, (ii) you reply with changing the deadline to a later, reasonable deadline. And you tell the sender you will wait for his approval of this postponement. Do not work on the task until you got the approval for your new deadline. To my regret, I have the experience that people always accept the new deadline. So you have to do the work anyway. But this time with less stress.

3.D.2.C.3 Extremely unreasonable requests

I know from experience that many organizations and their managers do not have their priorities and administration straightened out. The only way to solve that problem for them is to put their coworkers, clients and subordinates under stress.

There always will be an 'unexpected' visit from the American ambassador, or from a Minister of Education of China (China must be very big given the number of Ministers of Educations that are visiting the world).

There is always some middle-level manager that has forgotten about his duty. There always will be some unexpected pot of money and you have to put in a pre-proposal asap.

It all boils down to the fact that you will get an email, or fax, summoning you to send the same day, or the next day a couple of pages by email.

This email has cost the manager half a minute and requires hours of unexpected work on your side. There is only one remedy: plainly refuse to do the job. Teach them a lesson.

3.D.3. Multiple actions required

Getting an email that requires multiple actions on your side is a nightmare. In the first place, tell the sender to send only emails that contain one requested action per email the next time. He'd better send you five emails each containing a separate action, than one email containing five items.

You must put the separate actions separately on your To-Do list. For many scientists, their email box, with possibly some subfolders, *is* their To-Do list.

One possibility is to split the multiple-action email, by forwarding the email several times to yourself. You must edit the email in such a way that each forwarded mail contains just information about one single action.

Another possibility is to put several items one by one on your To-Do list (for instance with *Outlook*) and cut and paste the relevant parts of the original email into these notes, and then you treat them one by one.

3.D.3.A Action for more people

The worst that can happen to you, is that you are not the only recipient (more names in the *To*-header) and the message requires actions on the sides of more people. This ambiguity always creates trouble. Either all the recipients hope and expect that the other guys will do the work. Or two people do the same job.

If you got such an email, you must immediately straighten out the situation: just send an email to the sender with a cc to all other recipients and tell him that his email is unclear. And make obvious that you will not undertake any action until you get an email with only yourself in the *To*-header with the message body only containing the requested action you are supposed to carry out.

3.D.3.B Response forms

You will regularly receive response forms and surveys from newspapers, employers, science supporting agencies, companies, etc. Just never fill out these forms. Even if they phone you to remind you of their imposed deadline, tell them you won't do it.

The only thing that can happen to you when you do fill out request forms is that you get more of those forms. Those surveys are hardly ever going to be used for something useful. When all the items have been gathered and could be implemented, a new manager has taken over with totally new and 'fresh' ideas.

3.D.3.C Endless actions

Sometimes people have read one of your papers superficially and they want to start a scientific discussion with you. This is endless. It will never stop. Only embark on discussions with scientists that know what they are talking about.

Try to divert the start of a long series of time-consuming scientific discussions by pointing out the right textbooks and papers that first should be read. Get one of your colleagues to take over the discussion ("My colleague X at *MIT* knows much more about the subject", with a *Cc* to colleague X). If you want to delay your main competitor you can also refer people that want to start an eternal debate to this competitor (supply them with his email address).

3.E Dangerous attachments

Due to the many accidents that have happened with malicious attachments, and due to continuous press attention to virus proliferation, email users are getting more and more aware of the danger of receiving attachments.

Some of the obvious file extensions that are dangerous include: "exe", "bat", "scr", and "zip" (with password). A number of email clients do not even allow you to receive dangerous attachments.

3.E.1. Dangerous attachments

If you receive attachments having dangerous file extensions associated with them, you might have a problem. If you do not know the sender and/or do not expect such a file from the sender, just delete the whole email. If the file was of value, it is the responsibility of the sender to package that information in an acceptable wrapper. Just renaming the attached file on their side would solve the problem. Tell them so.

Sometimes you expect and get a file with a potentially dangerous file extension. (As always, you should have an anti-virus scanner scanning email attachments.) This situation may arise if you buy new software that is sent to you as an exe file. If your software client (like *Outlook*) does not allow you to receive the attachment, you have to invoke an old email client (old versions of *Outlook Express* for instance) that allows you to save the attachment separately.

A very stupid habit of people is to save files (documents) with a different extension than the default for the application (such as "doc" for *MS-Word,* "txt" for ASCII files). Or they save them with no extension at all. This non-standard naming is dangerous. You'd better refuse these files right away and tell your colleagues to use the standard file extension naming convention.

3.F Dangerous emails

Nowadays criminals try to send you emails to get hold of critical private information like credit card numbers and username/password combinations for your banking programs. They want to get to your bank account, *PayPal* account or other internet banking system.

There is a very safe way to be warned against those dangerous emails. The criminal emails are always in html format. Instruct your email client to show you the raw uri's, rather than the html version of the hyperlinks. Then you immediately see that the email has in it some dangerous hyperlinks.

The standard trick criminals use, is to use an internet alias very much like the banking system they want to rob. So you see "http://www.paypal.com" indeed, but then somewhere there is a period "." or some other suspicious character in the uri, followed by a link to *Perl* or *PhP* script (extensions like cgi, pl, or cgi) and then you should smell the danger.

https://www.paypal.com/

4 SENDING EMAILS

4.A There is no unsend

You cannot 'unsend' an email. That is the big problem. Mixing up the contents of the *Cc*-header and the *Bcc*-header can result in having made enemies forever. With some email clients you can postpone the actual sending of emails. This queuing can be very handy. The queued emails can be checked later, when you are less emotional for instance.

4.B Formatting of your message

Avoid sending styled text as much as possible.

So no accents, diacritics, ligatures etc. Sending styled text (rtf or even more proprietary) is against the philosophy of the internet. Even sending html files (web pages) should be avoided whenever possible.

4.B.1. ASCII

The crucial communication rule on the internet is:

"Send information as plain text (ASCII) whenever possible".

So do not use styled text with varying font size, italic, bold and colored fonts, and try to avoid sending attachments.

Many email clients allow the user to add formatting to their text. Popular formatting standards are html and rtf.

4.B.1.A ASCII styling

Even with plain ASCII there is still a lot you can do to structure your emails.

You can use uppercase and lowercase. Do not use uppercase in the middle of a sentence to indicate emphasized text: it will give the impression you are shouting. "You have to do this before MONDAY" is insulting. Be creative, like "You have to do this before *Monday*".

Underline text like this: "_This is very important_".

You can use strings of asterisks "*" as text separators. Software developers have to stick to plain ASCII when writing computer code. They are very creative in using all kinds of ASCII characters to have text comments stand out, like:

"////////////////////////////
////// RESULTS /////////
/////////////////////////////".

4.B.2. Rtf

In principle, the idea behind *Microsoft*'s rtf (*Rich* text format) is appealing: a general cross-platform encoded 7-bit ASCII. The problem is that *Microsoft* is continuously changing its proprietary rtf-standard, making it very difficult for independent software vendors to develop software with backward compatibility. In addition rtf-files tend to be bulky. You need a viewer (which many email clients have) to be able to read the formatted text.

If you insist on using different fonts in your email messages, like *Verdana* and *Garamond*, and font styles, like bold and italic, you cannot use plain text. My advice is to use html in that case, and not rtf.

http://en.wikipedia.org/wiki/Rich_Text_Format
http://en.wikipedia.org/wiki/Verdana
http://en.wikipedia.org/wiki/Garamond

4.B.3. Html

Html is a reasonably stable set of formatting rules. Html files are in ASCII, but it is encoded ASCII. You need a viewer (which all email clients have) to show the formatted text.

http://en.wikipedia.org/wiki/Html

4.B.3.A Figures

You can include figures as attachments. If, in contrast, you want to include figures directly somewhere in your email message, a number of problems arise. In html each figure constitutes a distinct file, separated from the text file. Your email client has to do all the bookkeeping when reading, sending or storing the embedded pictures. The email client of the receiver might do it differently from the sender's email client. Embedded figures are regularly ill-positioned and ill-scaled. So do not embed figures, just attach them.

4.B.3.B Lack of standardization

You are in the clear if you use very simple html formatting in your email. The best is to prepare your message outside your email client and copy/paste it. As soon as you use more sophisticated tricks, such as CSS files, you run into a minefield. It is the nightmare of all companies that do their marketing through email.

There is absolutely no standard. *Microsoft* suddenly decides to use *MS-Word* as html-engine in *Outlook* instead of *Internet Explorer*, with a lot of subtle changes. Many email servers strip advanced html tags they do not like.

An essential test on critical html email messages is to send them to yourself first.

http://en.wikipedia.org/wiki/Cascading_Style_Slides

4.B.4. Hyperlinks

If you include one or more hyperlinks that you want to be clickable by the receiver in your email message, you (or better your email client) must send the email message in html format. The reason being that only in html the formatting of a hyperlink is well described.

When the hyperlinks are an essential part of your email message, you must send the email first to yourself.

Only then can you check, by clicking them, whether the hyperlinks are correct and alive, or incorrect or dead. Sending an assertive email to the dean with a dead hyperlink would work very much against you, unless the dean does not know (quite possible) what a hyperlink is.

4.B.4.A Wrapping

Email clients have the bad habit of wrapping long words, even if you configure them no to do so. This is terrible if your long word is a hyperlink that should be clickable. If the hyperlink gets wrapped it has become close to useless. If you want to be sure that a long hyperlink survives the emailing process, attach it as a plain ASCII text file.

4.B.5. Hard returns

The use of hard returns (hitting the ENTER key of your keyboard while composing your email message) in emails is a constant source of irritation and confusion on the side of the receiver. Only use hard returns if you want to end a paragraph. If you introduce hard returns within a paragraph the receiver will get your message having wildly varying line lengths.

If you want to be able to compose (or read) your email in narrow columns, do not use hard returns, but make the window of your email client narrower.

Failing to follow this advice will result in emails with hard returns that are ugly to read (a lot of white space) and very cumbersome to cut and paste into word-processing programs.

The replacement of soft returns – for instance introduced because word wrapping is turned on – by hard returns and vice versa is an eternal source of cluttered emails. Receiving and sending clients only partly follow the rules opted for by the user. These erratic replacements, together with the only partly user-controllable format changes between plain ASCII and html, make the hard-return issue unsolvable. For the specialists there is an acceptable solution: enforce your email client to send your email messages always Base64 encoded.

4.B.6. Bad characters

Given the fact that all Internet protocols are voluntary conventions, you will regularly get emails by people violating these conventions, resulting in the appearance of unreadable strange characters. *Unix* freaks or users of strange code pages (the language system of the operating system) regularly cause this irritation when they send emails.

4.B.7. Font

Your email client allows you to prescribe the font you are using for the mail you send. It is tempting to use a fixed-pitch font (each character has exactly the same size, like in a type writer, also referred to as monospace), like *Courier, Fixedsys*, or *Andale Mono*. This is useful in columns, or when using numbers. However, there is no way you can force the receiver to also use a fixed-pitch font. If your email message indeed can be read much better with a fixed pitch font you should send it as an attached file made by a word processor.

Remember that sending font information in the header of the email message is not something the receiving client has to honor. Even if your email client gives all the font information in the message header, the email client of the receiver might do with it whatever it likes.

Keep in mind that emailing is about short email messages containing atomic information. Styled text is almost always a nuisance.

http://en.wikipedia.org/wiki/Courier
http://en.wikipedia.org/wiki/Fixedsys
http://en.wikipedia.org/wiki/Andale_Mono

4.C Requested action

If you require an action from the receiver, tell him immediately and explicitly at the beginning of the email and if possible also in the subject header:

Subject:
"Information about my problems with personnel department. No response expected"
Body:
"John,
This email concerns the recent problems with our personnel department. I just want you to keep informed. No action on your side is required. ..."

or:

Subject:
"Please talk to dean"

Body:
"John,
Below I will describe a clash I had with our institute director. Could you please intervene and talk to the dean. ..."

4.C.1. Single request only

Never ask multiple actions from an addressee. Write the person separate emails, each containing a well-defined separated requested task. The receiver will be grateful and will act upon your emails quicker. Put a deadline in your email. *Bcc* yourself and put that copy in your To-Do list. In that way you can react if the deadline has gone overdue.

4.C.2. Requested response time

Never expect people to react immediately. If they have studied this guide thoroughly, it will take quite some time before they answer you and that is good, for them and for you.

If they have to reply to you within one or two days, because otherwise you would get into great problems, you have your priorities wrong.

4.D Special emails

Some email messages belong to special categories, like confidential or formal. An email message can be sent very quickly. Often too quickly. With wrong entries in the *Cc*-header. Or with a sentence you thought you had, but in fact had not, deleted. These oversights can be very embarrassing when sending confidential or emotional messages.

4.D.1. Sensitive emails

Sensitive or otherwise important emails with a complicated *To-*, *Cc-* and *Bcc*-header, with hyperlinks or attachments should first be sent to yourself. Paste the full contents (not just aliases) of the *To-*, *Cc-* and *Bcc*-headers as text in the body of your message and erase the *To-*, *Cc-*, and *Bcc*-contents temporarily. Put your own name in the *To-h*eader. Preferably use one of your email accounts not part of the email account from which you are sending your letter. This is because some email programs, like *Outlook*, have the irritating habit of sending local emails always internally. After you have received the email, looked at the *To*, *Cc* and *Bcc*, opened all attachments and checked all hyperlinks by clicking them, you are able to send the email for real. Blind copy yourself, so to be able to control any damage immediately.

4.D.2. Formal emails

In a small number of cases (1%) it is useful to write your emails in a very formal way. It takes much more time to implement this advice, but it will help you to achieve your goals. Especially when communicating with managers.

There are a number of reasons for writing formal emails: (i) When you have a conflict with your management ("Why be a scientist if you can be his boss"). Some very high-up manager overseeing your escalating conflict at a later stage might want to intervene. The more formal and informative your emails were, the higher chances are that you win your fight. (ii) The receiver immediately feels the pressure when he receives a formal email. He knows that at a later point in time this email might show up in a dispute. So the receiver will indeed act, or act quicker.

4.D.2.A Letter simulation

Write the formal email in such a way that if it would be printed, rather than emailed, it could, without any further editing, be signed and put in an envelope and be sent by mail, or be sent by fax.

4.D.2.B Name and address information

Put the street address information in the email, as if you are typing a letter:

"To:" followed by title and professional capacity and exact postal information.

4.D.2.C Reference

Find some way to put a reference there. Something like "Reference 1233/AL" It tells the receiver immediately that you, the sender, will file this email in a professional way.

4.D.2.D Salutation

Always use the formal term of address, and if you know the addressee personally, in addition the informal "Dear director, dear John," is applicable.

4.D.2.E Recap

If your formal letter is part of an ongoing dispute, recap the history very shortly. You want the receiver (and possibly later moderators) to get the point immediately. Do not expect them to dig out your old emails.

4.D.2.F Data

If you have to put data in your formal email, like contract dates, amounts of money, proposal identifiers, get them right, get them exactly right. It proves you have, in contrast to your opponents, a complete and reliable filing system. You know your facts. Your managers are bound to make mistakes with these numbers. Correcting their mistakes strengthens your case considerably.

4.D.2.G Attachments

List in the body of the email the attachments as enclosures: "Encl.: ...".

4.D.2.H Formal cc

List at the end of the body explicitly the persons that are present in the *Cc*-Header. Use all formal titles and titles. Example:
"Cc: Prof. J. Johnson, director XYZ institute
Dr. L. Schwartz, chair of department"

4.D.2.I Formal bcc

Never use *Bcc* in a formal letter. If it later turns out that you have used blind copies you have weakened your position tremendously. You should fight with open arms.

4.D.2.J Example

"To:
Dr. W.G. Williams
Director Personnel Department
Univ. of X
123 Mountain Avenue
Ca #####,

Date: 12 February 2006
Reference: 120205/14.05/AL

Dear Dr Williams, dear Bill

[Body]

Sincerely Yours

Barry Y. Young
Professor of Physics

Cc: Dr. A.W. Gerald, director ABC institute,
Dr. G.K. Peterson, dean of sciences

Encl.: Letter of Agreement of the National Science Foundation"

4.D.2.K Fax has legal consequences

Formal emails carry no legal weight. People can always claim not to have received them. The same holds true for correspondence sent through regular (snail) mail. Registering mail is a hassle.

The fax is a real saver. In a number of countries, in legal disputes, sent faxes will be considered to be delivered. Even if this legal status is not rock solid, many organization will feel the pressure if they receive a fax.

4.D.3. Confidential emails

When writing confidential email, be careful. Before you know it, you have sent it (*Cc* or *Bcc*) to the wrong person(s). Minimize the number of confidential emails. Minimize the information in them. Write something like "I am in favor of the first candidate" rather than "I am in favor of candidate John Doe". Eavesdroppers and secretaries will hate you for this.

4.E Header

Basically each email message is just one big ASCII file. All the email client does, is parse the file and break down the information into header, message body, included figures, attachment, etc.

4.E.1. Automatic address

When you reply, redirect or forward an email, the client will (hopefully) just copy what it found in the received message header, like the raw address and possibly an alias.

4.E.2. Reuse of email addresses

The *To-*, *Cc-* and *Bcc*-headers take as entries email addresses. It is quite cumbersome to remember or cut and paste email addresses of recipients. For this purpose all clients offer some kind of address-book service, often with transparent auto-complete features.

4.E.2.A Address books

Address books can be very useful. So software vendors do their utmost to make the varying forms of address books mutually incompatible.

A terrible situation arises when the client, like *Outlook* in combination with *Exchange Server*, is programmed to find the email address through an extra level of indirection: the address book tells the client where (some database) it can find the email address, rather than telling it what the address is. This leads to disaster if you are managing your email from a location where this database is not accessible. Suddenly the addresses from your address book do not point to valid email addresses anymore.

> If the email client leaves you a choice, try to get the raw email addresses in the address book. That will always work.

4.E.3. Header part *To*

Never put more than one name in the *To*-header. If you do not follow this rule your emails will remain a continuous source of confusion and irritation.

In the event that you request action by more people, send separate emails, possibly with cc to the other people (that you originally planned to have in the *To*-header).

If you do not require action from any of the recipients, put them all in *Cc*- or the *Bcc*-headers and use a precise salutation to make clear what defines the addressees, like "Dear group members" or "Dear participants to the XYZ conference".

4.E.3.A Email to all

In many organizations distribution lists, a compilation of email addresses, are being used. Examples are "all@yourdomain.com" or "staff@youruniversity.edu"

Invariably the 'mail-to-all' email addresses are abused. I get irritated daily by receiving emails that some key has been found on the reception desk or that some car has its headlights on. Sent to 200 people.

Only amateurs like to send to all. It is a mere sign of immature behavior to send the send-to-all emails, and a sign of immature management to allow relaying to all. Information meant for all should be dispersed through bulletin boards or public folders.

4.E.4. Header part *Cc*

In the *Cc*-header you put addresses of people you want to inform about the email you have sent to one person in the *To*-header. Never request an action from somebody you sent the email cc. Do not bother people too much with cc-ed emails.

Meticulously chosen *Cc*-recipients can help a lot to speed up the action you have requested from the *To*-recipient. It puts pressure on the *To*-recipient. This can easily run out of hand, because in many cases the *To*-recipient will interpret your *Cc*-action as a motion of distrust.

4.E.5. Header part *Bcc*

There is no way the *To*-recipient can know who will get a blind copy. If he later finds out, because one of the *Bcc*-recipients exposes himself, you have maneuvered yourself into an awkward position.

Be very, I really mean very, cautious with putting people in the *Bcc*-header.

4.E.6. Subject

Indicate the subject of the message by a text in the subject field that is as short and as clear as possible.

If the subject is well worded, the receiver can already guess the content of the email.

When you forward emails, change the subject. Subject: "Visit by Dr. Wang, applicant for postdoc position in group X" is much more informative than "Application".

Do not use the "re" option.
Never name more than one subject. Write different emails instead.

4.E.7. Attachments

In all cases try to avoid sending attachments. Put the information directly in the email as plain ASCII. I hate emails that have as body only one sentence, saying: "See attachment".

4.E.7.A Reason for attachment

Always say in the email message that you have attached one or more files, and tell the reader what is in the attached file. Do not let them guess. I notice that I regularly forget to open attachments if I have not been warned about their presence in the email.

4.E.7.B Stupid attachments

Sending emails with *MS-Word* attachments that contain plain text information, which could have been sent easily directly in the body of the email, is just plain stupid. But for some reasons many secretaries like to do it anyway.

4.E.7.C Color

If your attachment uses colored fonts or colored pictures, it means that it should be printed with a color printer. Tell the reader in the body of the email that the attachment should be printed with a color printer. Readers often print attachments by default on a black-and-white printer without ever looking in detail at the content. Furthermore, you should prepare documents that have color information in such a way, that they are still easy to read when printed on a black-and-white printer.

4.E.7.D Naming of file

The name of the file you attach should have an extension that exactly reflects the contents: doc for *MS Word* and pdf for pdf-files, etc.

As chairman of hiring committees for full professors I have received emails from a number of applicants that all contain the attachment named "cv.doc" of "cv.pdf". Please use unique names that mean something to the receiver. Always use English names. For instance: "j_logan_cv.doc". Attachments with such names can be saved by the receiver without the necessity of renaming them.

Do not use spaces in names. File names with spaces are not recognized by a number of *Windows* command-line programs (unless you use awkward quotation marks). Use underscores ("_") to separate text parts of a name.

Do not use filenames with mixed case. The *Windows* operating system, unlike the *Unix*-family, is not case-sensitive.

4.E.7.E Business card

Scientists receive tens of emails per day. Many of them without, and a few with, attachments. If there are files attached, the email client flags this situation to the receiver. The receiver sees this signal and he knows that he probably has to open it. That is to say if the attachment is a necessary attachment.

The habit that some people send emails with their business card attached by default, is very unfortunate. The recipient continuously thinks that there is a necessary attachment, whereas it contains a redundant business card. Put the business-card type information at the end of an email. For instance by using a signature file. Signatures in email sent to colleagues that receive an email per day from you on average, should not have the full address information. Leave the "Prof.dr." stuff to the Germans.

4.F Body

The email body is defined as the text of the email message without headers and attachments.

4.F.1. Style

Humor in emails is bound to be misunderstood.

> Avoid sarcasm; as a matter of fact avoid any strong emotions.

In many cases the receiver, or his secretary, has a mother tongue different from English. American humor is different from British humor. Just write short, businesslike texts without any spelling error.

4.F.2. Salutation

Any email should have a salutation. Many emailers do not realize how important precise salutations are.

A number of bad examples: "Hi there", "Dear All", "Hello". A receiver, reading the first line, does not know at all why he gets this email. Is he in the *To*-header, or in the *Cc*-header? Are there more persons in the *To*-header? The receiver has to guess, because you have been lazy.

Good examples are "John", receiver William immediately realizes that he is in the *Cc*-header or in the *Bcc*-header, and he very likely does not have to take any action on the email. Another good example: "To all WZI faculty members". A secretary that receives the email knows immediately she is in the *Cc* or *Bcc* department. And all faculty members know that the email is meant for them.

4.F.2.A Friends

If you know the receiver very well, do not use "Dear" and the likes. They distract. The relevant information should be placed as left as possible. So rather use "Bill" than "Dear Bill".

4.F.2.B No salutation

Some of my colleagues are so busy that they do not have the time to put a salutation at the beginning of their email messages and replies. This is insulting and inefficient for the receiver. Insulting, because the sender makes clear that the receiver does not deserve much of his time, and inefficient, because the receiver needs much more time to classify the email.

4.F.3. Requested action

If you write a somewhat longer email in which you request an action from the addressee, put the request right at the beginning:

"John,

In this email I ask you to write ...", and if possible in the *Subject* header.

4.F.4. Trivial body

Very short email messages can often be put fully in the *Subject*-header. Like subject: "I am going home now". Better of course is "Henry D. is going home now". Real smart-asses put in the message body: "See Subject". This is inefficient. Exactly copy the contents of the *Subject*-header in the body.

Never leave the body empty.

4.F.5. Signature files

A signature file contains a number of text lines containing your name, affiliation and contact information. Your email client can be configured to append these lines automatically to all your email messages. Signature files can be very convenient. But often they are much too long and contain TMI (too much information).

Sending your local colleagues your phone numbers, all your affiliations, all your titles and more stuff is irritating. In addition it is showing a sense of insecurity about yourself.

So have several signature files, formal, and for colleagues, in two languages (English + native language).

If you ask somebody to phone you, or you expect somebody might want to phone you, put your phone number at the end. That is much more convenient for the receiver than his going through his contact folder (which is very likely outdated).

4.F.6. Grammar and spelling

Never introduce deliberate spelling errors. It just proves that you are lazy. Days of the week should be written starting with a capital. And "i have seen ..." proves your laziness and disinterest in good communication. By using "i have", the email writer spares himself the use of the shift key. However, the burden is now on the reader. Wrong spelling makes the email much more difficult to read. In addition, you give the receiver the impression that you do not give a damn about his time.

Emails are often considered to be a form of informal communication. This is a wrong perception, as email communications are not volatile. They will last forever.

4.F.6.A Date

Never use a date format that contains only numbers, like 10-11-2006. This means different things in different countries and it requires a lot of calculation on the side of the receiver. Be as verbose as possible about the date: "Friday, 06 October 2006". If you ask somebody to be present at a meeting in this way, he might know immediately that he cannot attend, without ever consulting his calendar (which could involve switching his computer on), because he always has a group meeting on Fridays.

4.F.6.A.1 Relative dates

Never use relative dates like "next week" or "next Friday". This is always ambiguous.

Use a verbose date format: "On Friday 11 June 2005 we will ..."). This is much more informative than "On June, 11 we will ...".

Use of "Tomorrow" or "Today" is confusing. If you want to use them anyway, combine with the day, like "Tomorrow, Thursday, 12 July".

4.F.6.B Time

Anglo-Saxon culture uses "pm (PM)" and "am (AM)" whereas most continental Europeans use the 24 hrs time format. Scientists should use the 24 hrs system, as there will never be an ambiguity.

4.F.7. Figures

People that put figures in the body of emails, do not understand the idea of emails. Flat information, that is what it is all about. Figures are often displayed in a different size than meant by the sender. If you need to send a figure, or figures, attach them with sensible names. To use figures in html is a well-known trick for spammers to test the validity of your email address.

Hyperlinking figures to internet addresses in your emails is just plain dumb. The receiver will need to be on-line to be able to read your email. Furthermore, email clients tend to disallow this, because spammers use this method very often.

4.F.8. Tables

If you use formatted tables in your email, you must be out of your mind.

4.F.9. Planning a group event

You will probably often get cumbersome lists in an attached file with possible dates for an upcoming event. If you write such a list yourself, it is bad practice to put this list in an attached file (this means the receiver has to open the attachment with another program). Just place it in the body of the email. Design it in such a way that the addressee can return the email with minimal effort.

Just put the date list in the following format (and train your secretaries to do so as well):

"Y/D/N Morning   Fri 6 June
Y/D/N Afternoon   Fri 6 June
Y/D/N Morning   Mon 12 August

Y=Yes, D=Difficult (optional use of D), N=No".

Look at the alignment. The receiver has a number of starting points when reading this list. He only has to delete four characters per line and can return the email.

Planning a group event with a scheduling program like *Outlook* is only for managers, not for professionals.

4.G Replying to emails

Replying to a received email message with a new email message has as main advantage that the full text of the incoming message is ready to be used as part of the reply.

> Instruct your email client not to put foolish characters like ">" or ">>" in front of each line of the original message.

Also, emphasizing the difference between the old text (of the received message) and the new, by using different colors, is a bad idea. Many of these markings do not survive the emailing process, and in addition make it more difficult for any receiver of your email to copy/paste your text without extensive editing.

4.G.1. Check the end of your message

During composition of an email message the text structure might become quite complex. You might have pasted text. You might have appended earlier messages.

When you are about to send an email message, you will usually not see the whole text of your message in your viewer. You think you know what is in the message, but you might well be mistaken. For instance, because you forgot that there are still some trailing lines, invisible in your window because of a number of blank lines preceding them.

Place the text cursor right at, or after, what you think is the last character of your message, often your name at the end. Press the DELETE key continuously, or select all the data after your name and delete. Only then can you be sure that your name actually marks the end of the message. If you forget this procedure, you will regularly send unwanted trailing text along with your message; with possibly embarrassing content.

4.G.2. Send your own format

Customize your email client to send plain ASCII, even if you reply to received messages that contain styled text. If you see that your client is about to send your reply in non-ASCII, change the format (in *Outlook:* select *Format* from the *Standard* toolbar)

4.G.3. Standard reply

If some people practice bad email policies, especially if they are associated with administrative departments, you will have to teach them a lesson. Your time is valuable, and they are just there to support you. Use a standard stationary (signature file) with the following checkable options (edit and extend according to your own wishes):

"Dear Sir,

I will not act on your email for one, or more of the following reasons:

1. [#] You sent it to an obsolete address
2. [#] Contained unacceptable attachments
3. [#] Promised attachment is missing
4. [#] Text is not sent as text, but as a scanned image
5. [#] Attachment contained simple text that should have been sent in the body
6. [#] Body of the email was empty
7. [#] Imposed deadline is unreasonable

YS
Donald Duck"

You can check those reasons that apply.

4.G.4. Trimming

Try to keep emails as short as possible. When replying, it is never a good idea to include the whole text of the email you are replying to. In a few iterations the email contains hundreds of lines prefixed with multiple characters (like ">", ">>"). Together with inappropriate hard returns, the whole email becomes absolutely unreadable.

If you summarize the received email in one or two lines yourself, there is no need to include the email.

Subject:

"Unable to join program committee"

Body:

"John,

Thank you for your recent email concerning the organization of a conference on biopolymers. You asked for my participation in the program committee.

I think the conference is important and timely. I will advise my graduate students to participate. I feel honored, but unfortunately I cannot accept your invitation. I have another conflicting obligation that I agreed on a long time ago. I cannot be at the conference and it makes no sense to be on a program committee without being able to attend.

I wish you all success with the conference.

Greetings
Donald Duck"

In this way you did not need to include the email.

4.G.5. Reply to all

Replying to all should be strongly discouraged. Leave it to the old guys that just learnt to handle a keyboard.

4.G.6. Waiting time before you reply

Never ever answer an email immediately. It will always bring you more work. Delay and stall. People should realize that you read your email at most once per day and send your emails only after a week or so.

Really cool are email clients that let you delay the sending of an email. You just prepare your email and schedule it to be sent in a week or so. It can be taken from your To-Do list. Blind copy yourself.

4.G.7. Out-of-office reply

Automated replies are for amateurs. They are dangerous for a number of reasons.

Firstly, they are sent immediately. So they give the impression to the receiver that you are not in your office, but that when you are, you have the habit of answering your emails instantaneously.

Secondly, email clients have lots of problems (bugs) with respect to sending out-office-replies to spammers. Your spammers get information that your address exists and that is exactly the information they like, they will use, and they will sell.

Thirdly, a number of people, like academic administrators, should not know about your whereabouts.

Fourthly, criminal spammers that send you an innocent email will know that you are away from your office and probably also your house. This could give them wrong ideas.

4.H Redirecting

Redirecting is quite different from forwarding. Redirection occurs when you have several email accounts and want to receive them in one inbox.

For the receiver the redirection is transparent (only deep in the message headers can the redirection be traced).

4.I Forwarding

4.I.1. To one person

The more of your received email messages you can forward the better. Make very clear to the forwarding recipient that you do this for (i) information only, for (ii) information and you expect an action from his side. Make clear what action you expect. And also make clear that you have informed the original sender that the forwarded recipient will be responsible for (that) part of the action. By doing so, reminders on the action not being completed yet, will not go through you.

Do not change the subject. Just add a few lines at the top of the message body to indicate the expected action.

If you do not put any private information in the top of the forwarded message, you can cc the forwarded message to the original sender. In this way you have written yourself out of the loop with just one email. Otherwise send the original sender a different email with the information about the guy that will be responsible for the action. And you are off the hook once again.

4.I.2. To more people

Forwarding to more people must mean you do not expect any action of the addressees. Otherwise you make a mess out of your emails.

4.J Credit card

These days, criminal emailers become cleverer by the day. It seems stupid to have to say this here:

> But never ever send credit card details by email.

Either fax or use only websites of those companies that supports SSL (Secure Socket Layer, you will see an "https" rather than an "http"), a very safe encrypted way of sending information over the net. Conference organizers are often amateurs and require credit card details by email. Just plainly refuse.

The same holds for all other information like passwords.

http://en.wikipedia.org/wiki/Secure_Sockets_Layer

5 MANAGING ACCOUNT(S)

Managing email accounts can be very tiresome. You would like to have a situation in which, wherever you are in the world, you are able to receive your email, send your email and have all your archived emails at your disposal. And in addition you want these new activities to be archived as if you have been performing all these actions from your local pc. Achieving this goal is far from simple.

5.A Multiple accounts

Nowadays many scientists will have multiple accounts. Either because they have more than one affiliation, or they want to maintain low-security mail accounts – like *Yahoo!*, *Hotmail*, or *Google* – and a high-security account for their professional correspondence.

http://www.yahoo.com/
http://www.hotmail.com
http://mail.google.com/mail/help/intl/en/about.html

5.A.1. Prestige of account name

Be aware that your email address is in some way your business card. So when you work at *Harvard University* and write an important email to the dean, it does not make a good impression to do it through your *Yahoo!* account.

You can more or less hide your low-profile email address when you send messages through it, by supplying an alias inside your sending email address, like

"J.Doe@yahoo.com"<John Doe> and instruct your email client to use as the reply address your prestigious mail account, like "J.Doe@harvard.edu.org", the latter of course without an alias.

5.A.2. Automatic forwarding

When you have multiple accounts, you may want to choose that emails received by some of your mailboxes be forwarded (redirected is a better term) to one of your other mailboxes.

Whatever way you choose to have your email forwarded from one (POP3) server to a second server, your email client – or the first server itself – should be configured such as not to leave a copy on the server after it has forwarded the message. If you do not configure your email client in this way you will continuously end up with numerous duplications of email messages: a horrible situation when your inboxes are your To-Do lists.

A drawback of not leaving a copy on the server is that you must be very careful when synchronizing your email, otherwise you will lose email messages.

5.A.2.A Hard forward

One solution to the problem of managing multiple accounts is having all the mail forwarded to one account. This forwarding can be set at a pretty low level at the server (by an administrator) and will be transparent for the user (you). If you choose for this option, have this forwarding set at the lowest possible software level of the server, with the least possible change of the message header. Any software problems later down the line will not affect the forwarding procedure.

The problem of course with low-level forwarding is that all the accounts get mixed up. If you keep multiple accounts for separate bookkeeping or archiving reasons, this hard forwarding is no option. Although, you could still separate the email messages from your various accounts later, if the client you are using, has powerful filtering capabilities.

The main reason for maintaining multiple accounts with hard forwards is that you want to have more SMTP servers available, so that you can still send your email if one of your servers is down. And a server that is down regularly, is a fact of life.

5.A.2.B Soft forward

Setting the forward to another account at the email-client level, with the use of filters on incoming messages, is not very convenient. The forwarding will only apply if you are logged on to your email account. In addition message headers are usually changed in such a way that the original sender gets lost in the *To*-header.

5.A.3. Multiple POP3/SMTP servers

As explained earlier, one reason for having multiple accounts is that when your server is down (this occurs regularly at any place) you can still send email through one of your other email accounts (with their own SMTP server).

Another reason is that some providers protect their SMTP server in such a way that you cannot use it outside their network unless you have a VPN connection. It is convenient to be able to send email messages without having to be connected to a VPN server.

SMTP servers have to be protected, as otherwise any spammer could send a million messages to the unprotected SMTP server and tell it to relay them around the world. There are basically two ways to enforce this protection: an IP-number check or a username-password combination. Or possibly a combination of the two.

Try to have at least one SMTP account at your disposal that does not require an IP-number check.

If you do not have this extra SMTP server and if you are not logged on to the local area network of your provider/employer, you will always have to make a troublesome VPN connection before you can send your email.

5.A.4. *Exchange Server* plus POP3/SMTP

The combination of *Outlook* running through an *Exchange Server (ES)* with additional POP3/SMTP accounts can be a nightmare. Nevertheless, this is a situation that many scientists (including me) will have to cope with. The biggest problem occurs when you are working with an *ES* that is on-line. Working off-line with *ES* in cached mode is what you have to do as much as possible. In off-line mode you are not sensitive to any network problem or *ES* problems, which result regularly in a computer that is slowed down tremendously. *Outlook* in combination with *ES* is absolutely horrible in the sense that it monopolizes your computer if the connections are slow.

In principle, when you are working off-line with *ES*, but you are connected to the internet, you can still receive email from your POP3 servers and you can still send emails with your SMTP servers.

Connect to the *ES* twice per day: in the morning when you come in, and in the late afternoon when you leave your office.

5.A.4.A Sending emails

When you are working off-line with *ES*, *Outlook* is very stubborn in choosing the SMTP server you want to use when replying to email. It will by default use the *Exchange Server*, but that is unavailable as you are working in off-line mode. You will end up with email messages sitting forever in your outbox. Or you end up with an *Outlook* instance that keeps on trying to connect to the *ES*. You have to manually change the SMTP server that should be used for sending (click on *Account* and choose one of your SMTP servers). And you have to do this for each and every email you send.

5.A.4.B Filtering emails

A big problem with *Outlook* in combination with *ES* and additional POP3/SMTP servers is that the filtering of incoming messages that do not arrive through the *ES* will not be done regularly.

5.A.5. Multiple Exchange Servers

Microsoft's Outlook does not allow you to have more than one *Exchange Server* working at the same time. If you have to work with two *Exchange Servers* (as I have to) you can have the luck that one of the *Exchange Servers* also exhibits POP3/IMAP and SMTP functionality. In such a case you can combine them in one *Mail Profile*. Otherwise you have to use two *Profiles* with two inboxes (horrible). In the latter case setting a hard forward on one of the two *Exchange Servers* is the preferable solution.

Outlook really messes up when you have more SMTP servers. It seems to know better than you what SMTP server to use. Terrible. Always check *Account* to see what server it will use when you are about to send a message.

5.B Synchronization issues

Scientists travel a lot. They write and send emails wherever they are in the world. They work on their papers and presentations wherever they are in the world. This high mobility requires quite some discipline as otherwise valuable information will get lost.

Here I only deal with the syncing of emails. But some of the solutions I recommend are also applicable to data files.

5.B.1. No syncing required

Your need for synchronization depends very much on how mobile you are.

5.B.1.A Fully Localized

If you have only one email account and only one computer on which you handle all your email activities, there are no synchronization problems. You seem to be in luck, but you are very vulnerable, because you have no backup, unless you have taken special precautions.

5.B.1.B One portable computer only

If you have only one computer but you travel a lot, this one computer will be a portable computer. If you do not use a network drive and download all your email to this laptop, there is no synchronization problem. In such a case all your received email messages will be physically residing on the hard disk of your (portable) computer. You are very vulnerable, because you have no backup, unless you have taken special precautions.

5.B.1.C Web interface only

Another situation where you have no synchronization problems is when you have only one email account that you handle only through a webmail interface (I assume that your webmail provider allows you to store large amounts of data on their server, which for instance *Google's Gmail* does.) The disadvantage is that you always have to be on-line to handle your new email messages or to search though your old email messages.

Although webmail interfaces get better all the time they will always have major disadvantages compared to performance of a locally running email client. Basically web interfaces lack speed and functionality.

http://mail.google.com/mail/help/intl/en/about.html

5.B.1.D Network drive only

If you download your email always to your network drive, wherever you are, there is no syncing problem. But you have lots of other problems, because your network drive must always be available. That is to say, you must always be connected to the internet and you must always use a VPN connection if you want to manage your email activities.

Your provider will probably backup the network drive. That means restoration of lost files will always need the support of the ICT-people of your employer. That is, unless they use a sophisticated RAID disk system, in which case you can do the restoration yourself.

http://en.wikipedia.org/wiki/RAID

5.B.2. Syncing required

Except for the cases mentioned above, you will have to worry about synchronization. For instance when you work with more than one computer and you want to store email information on your computers.

Due to the occurrence of many internet server problems, I can hardly think of a situation where scientists are not confronted with synchronization issues.

5.B.2.A Slow servers are a fact of life

If your original, naive, idea was to always work with an internet connection, so as to avoid any cumbersome synchronization procedures, you will soon wake up out of this dream. On many occasions your network connection will become too slow. Either because your network drive is slow or because you use an email client like *Outlook* that keeps on sending large amounts of superfluous data to and from the server.

5.B.2.B VPN

Nowadays VPN (Virtual Private Network) servers are becoming very popular. As such they became the target for many hackers and crackers. As a result, VPN servers are getting more and more complex, restrictive and they slow your computer down. For instance the VPN client from *Cisco*, required by one of my employers, cripples a major part of my computer's network activity.

If you are lucky you have a(n) (A)DSL or cable connection at home endorsed by your employer. Then the security is often an IP-range check and you will not need a VPN server.

http://en.wikipedia.org/wiki/VPN

5.B.2.C Synchronization challenges

The three major challenges in managing multiple email accounts are (i) not to lose any email messages, (ii) not having to deal with duplicated email messages all the time, and (iii) having available all your old email activities.

Losing recent emails, temporarily or permanently, can occur if you download an email from your POP3 server to a local computer (at your work or at home) and you do not transfer that message to a movable storage medium (like USB disk or stick) or do not copy it to your network drive. If you are that sloppy, your (local) inboxes will be different on your various computers: a horrible situation.

5.B.3. Movable storage

If you have no network disk, or do not want to use it (as in my case), you are stuck to discipline and physically bringing your data with you all of the time.

The discipline is three-fold: (i) wherever you go, you have to bring the storage device, (ii) you always have to make a local backup before you disconnect the storage device from your computer, and (iii) you have to protect the data on your storage device.

Not withstanding the discipline requirements, this approach is my favorite.
I can work on my emails on all my computers without any of them being connected to the internet. Only when I want to physically send or receive emails I need to have an internet connection. For this I only have to be connected for a short period of time; usually only twice per day.

5.B.3.A USB device

The best solution is to have a large capacity, fast and very compact USB 2.0 drive. A USB 2.0 stick might also do it. But a disk is much faster and has a much higher capacity. I have a light-weight 20 Gbyte *Arc Disk* disk. It contains essentially all my data of the last twenty years.

http://www.archos.com/

5.B.3.B Synchronization

You will also need a good synchronization program, so that all the files on your computer are in sync with the files on your USB 2.0 device. It requires discipline. Whenever you leave work (home) you have to sync and whenever you start at home (work) you have to sync again. That is four times per day. I use *Beyond Compare*. Synchronization of a few hundred Megabytes (due to a clumsy implementation of *Microsoft's Outlook* pst files, I have to synchronize a number of files that haven't changed) takes about a minute.

http://www.scootersoftware.com/

5.C Retrieving emails

When you put mail on your network drive or when you have the discipline to always synchronize (I have done this four times a day for years already) you can customize the POP3 (or IMAP) server not to leave the mail on the server.

However, you might end up with duplicates. Because to your program at home, the mail on your server looks new, although you have already downloaded it. Some email clients have clever tricks to find out if you have already downloaded the mail. But these tricks are never failsafe. Then you end up with duplicates.

For many of the popular email clients like *Outlook* there are third-party (usually not freeware) plug-ins to remove duplicate emails. I use *Duplicate Email Remover.*

http://www.mapilab.com/outlook/duplicate_remover/

5.C.1. Received mail

Notwithstanding the fact that you are very careful and very disciplined, and chances are that you will hardly ever lose a received email message, still it will occur.

Another pain in the neck is your colleague that is standing next to your desk asking about his recent email. You did receive it, but you have already filed it somewhere in your huge hierarchical To-Do List with multiple folders. It would take some time to retrieve it. In the mean time your colleague is making fun of your bookkeeping system. There is a solution to both problems: Inbox copy.

5.C.1.V.1 Inbox copy

Many email programs allow you to perform automatic actions on incoming mail. What you should do is the following. Make an extra folder on your server (possible with IMAP and *Exchange Server*), or in your local inbox folder. After your email has gone through all kinds of spam filters (that you might have installed there yourself), you save a copy of each and every incoming email to this new folder. Have your email program automatically clean this box by deleting old items (let us say older than two or three weeks).

In this way you have a very quick and handy safety net for all your incoming emails. In cases of emergency you can always retrieve all your incoming messages of the last (two or three) weeks.

5.C.2. Reply address

When managing multiple accounts, it might be worthwhile to instruct your email client(s) to send with the message a reply address different from the sender's address.

5.D Sending email

Sending email with multiple accounts can be a nightmare. (With *Microsoft's Outlook* it is a disaster.) Your email client has several options when instructed to send an email message, as it has several SMTP servers to choose from.

In my mind the best solution would be to allow the user to pick out the sequence of priority and if one mail server fails the next should be tried.

In such a case your mail would always be sent.

This is not the case with *Outlook* and possibly other programs. The program wants to figure out itself what is the best SMTP server. No doubt it often thinks this is the *Exchange Server*. But if you're off-line with your *Exchange Server* this does not work and your mail will not be sent. You have to manually assign a different SMTP server to your outgoing mail and request its sending.

Outlook regularly picks out an SMTP server that cannot be used because, from your IP-location, the IP-check fails. Again your mail will not be sent. Change the SMTP server that is associated with the email (click *Account*).

6 ARCHIVING EMAILS

Emails you have received or sent in the past, may contain vital information. Being able to quickly retrieve data present in old messages will be crucial for your functioning.

This email archive should be arranged in a set of hierarchical folders. If you devise a well thought-out structure, that structure can be kept year after year. *Microsoft Outlook* outperforms all other email clients in this respect. The flat structure offered by *Google GMail* does not provide this necessary structure, however fast their email search machine is claimed to be.

6.A Keeping old emails

Given the very low price of disk space these days there is no need whatsoever to delete any old emails out of economy reasons. You can – and should – keep all your old emails. Organize them in hierarchical folders. Many modern email clients offer this facility.

6.B Archiving formats

There is no standard file format for the archiving of email messages. The best type depends very much on the way your email client saves your email messages.

6.B.1. Ideal format

The ideal saving format – at least in my opinion – would be to have for each email message a separate directory containing the full email message (including all message headers, but stripped from its attachments) as a single ASCII file and in addition containing the attached files accompanying it as separate files.

My ideal has as a possible problem that it might be somewhat slower to find a particular old email message or one of its attachments. But the good thing is that the email archive remains readable forever and the readability is not connected to any proprietary file format that might become obsolete, of some commercial email client. *Eudora* (an email client that is sadly no longer being developed) has a way of saving messages that comes a long way towards my ideal.

6.B.2. Proprietary file formats

Email messages, including their attachments, are sent around the world as ASCII files. The first thing commercial companies do when developing email software is to turn these universally readable ASCII email messages into a proprietary object file format that can only be read and maintained with proprietary software from this one and only company. It is capitalism, all the way.

6.B.3. *Outlook* format

An extreme form of using a proprietary file format is the way *Outlook* does it. It keeps all the email messages, together with their attachments, in one file (a pst file, pst stands for *personal store*).

There are many disadvantages to this system. I will describe them in some detail. Many of the disadvantages hold for other programs with similar philosophies.

6.B.3.A Pst files

One drawback with the *Outlook* system is that you need *Outlook* to read and maintain your old emails. Who knows what will happen in ten years from now. *Microsoft* has already recently introduced a new pst format; to force you to regularly buy new software versions of *Outlook*.

In the future, old versions of *Outlook* might not be able to read new pst files and vice versa. Then it does not matter that you have them backed up nicely. Unless you keep on buying the new software versions of *Outlook*, your backed-up emails will become out of reach.

The pst file has a complicated internal structure. This had a lot of disadvantages. First you need the program itself to read the files. Furthermore it is very vulnerable. One mistake in the structure can be enough to destroy the whole file (with years of emails in it). For this reason *Microsoft* supplies repair programs for pst files.

Any slight change, as a matter of fact even reading it, changes the pst file. As a result the whole file (easily 200 Megabytes) has to be synchronized and backed up again. Horrible.

Given their vulnerability, you have to backup the pst files regularly. And frequently you should convert them into ASCII. I have to live with *Outlook*, but to be on the safe side I often use *Eudora* to import my pst files and convert them into readable ASCII files. The disadvantage of this is that the conversion is not full proof.

6.B.3.A.1 Two versions of *Outlook*

Keeping an old version of *Outlook* together with a new version is not an option for beginners. In the first place the old version will probably mess up your file associations (when clicking on a pst or msg file, the old version opens these files rather than the new version).

Keeping two versions of major applications, like *Outlook*, is dangerous as it might mess up the common files – usually called dll's.

It gives rise to what developers call *Windows* 'dll-hell'. Dll's are called daemons in *Unix*. Keeping two versions of *Outlook* is disabled in *Outlook 2007*.

6.C Filing system

The more attention you give to the filing system of your messages the better you can retrieve them later. But it requires discipline. Think deep about the folder structure.

6.C.1. Sent emails

For some psychological reason people take much more care of archiving received emails than of retaining old sent emails. However, it is essential to keep copies of all sent emails. Archive sent emails in folders, with names denoting the received month (like "03_February", the "03" ensuring the folders will be displayed in chronological order and the leading "0" ensuring that all folders names are vertically aligned).

Do not classify your sent emails according to subject. This is very time consuming and not worth the trouble.

6.C.1.A.1 Attachments

If you sent an email with an attached document, you must have a copy of that document in your file system.

If your email client keeps in its records a separate copy of the attached file, your sent archive will easily clog up (easily hundreds of Megabytes that you have to synchronize daily). I think you should regularly strip the attachments from the sent emails.

6.C.2. Received emails

File only those received messages that contain real information. Not the "it is coffee-break" stuff.

6.C.2.A Attachments

Different programs have different philosophies about what to do with received attachments. Some programs put them in separate folders. Other programs keep them with the email.

6.C.2.A.1 Saving with attachments

Programs like *Outlook* keep the received attachments with the email. This is principally the best solution as they constitute one unit. If you file your emails in a systematic way there is no need to save the attachments separately. You can move your email around in your email-filing system and the client will always be able to trace where it put the attachments of a particular email.

6.C.2.A.2 Saving separately

If you want to save the attachment somewhere in your file system (like in "d:\data\docs\", you easily end up with multiple copies of the same (usually large) file. In this case you had better delete the attachments from the email after you have saved it separately.

The location where you saved the attachment must be very logical. Otherwise later you will run into great difficulties when you want to read the old email again and cannot locate its attachment anymore.

It is much better to have a system where you do not need to save the attached file separately. Just make the incoming emails, with attachments, part of your data filing system.

Documents without accompanying email can best be filed by sending them to yourself as an attachment.

6.C.2.A.3 Version control

The problem with keeping all the attachments with the received email messages, is that you might end up with multiple versions of the same manuscript. Just file them together in one folder and delete the attachments of all but the latest version when the manuscript has reached its final form.

6.D Finding

It is a challenge to retrieve an old email out of a collection of thousands of emails. But often you will need to do just that. Many email clients allow you to search through (old) messages using a search wizard with search masks (date, subject, etc). They are useful, but often a little crude and impractical (certainly with *Outlook*).

You could consider using free indexing software (like *Google Desktop*). Or better use a superior, non-free, product, like *dtSearch Desktop*. This program indexes your whole computer, on a scheduled time, requested by you. With superb finding filters you can find anything on your computer. *dtSearch Desktop* can search through *Outlook* folders.

The downside of using indexing programs is that you have to do the indexing regularly, preferably once per day. These indexing applications monopolize your computer, so you'd better do it at unproductive times (idle time of computer or at night).

http://desktop.google.com/en/
http://www.dtsearch.com/

6.E Backing up

If you have a network drive, your data will be backed up by your employer.

If you use a large-capacity USB device as your primary data source, there is no reason for extra backup. You always have at least three versions of your data: on the computer at work, on your computer at home and on your USB device.

Regularly, for instance yearly, 'freeze' old folders in your email-filing system.

That is to say do not add any new information to them. Just make new folders. If your client allows it (*Outlook* doesn't) make them read-only.

7 SECURITY

A major part of this section on security consists of technical details.

7.A ICT group

The ICT group (any organization over 20 employees has a department of this kind, albeit with a varying name) is a bunch of professionals with autistic character properties by definition. That is why they are so good at working with computers and so bad at communicating with you.

Members of the ICT group love to make clear to you that they have administrator rights and you do not.

If you know what you are doing, you should have administrator rights on your own computer.

7.A.1. Are you a computer professional?

Here are the tests: Do you know the difference between UDP and TCP? Do you know what *IPv6* is? Do you know what an RPC is? If you know the answer to these three questions you should request being an administrator on your own computer (not on the network). The ICT people will discourage this by telling you they want to protect you from yourself. In many cases you'd rather be protected from them. Insist on getting administrative rights.

7.B Encryption

Encryption is not useful for your daily email messages. There is a number of neat encryption methods, like PGP (*Pretty Good Privacy*). But these methods are only practical if the communication is between a very homogeneous group of people. Scientists communicate with a heterogeneous group, consisting of colleagues, students, managers, secretaries, editors, reporters and so on. Many of them are not willing, or not able, to incorporate an encryption method in their email communication.

http://www.pgpi.org/

7.C Privacy

Realize that your employer and people mandated by your employer have the right to read your email. Moreover, on the internet nothing is anonymous. Web servers and email servers keep detailed log files, often for years. Nowadays storage capacities of hard disks, combined with sophisticated compression techniques, alleviate the necessity to delete log files or archive files.

Your network drive and even the hard drives of your office computer can also be inspected by your computer administrators.

7.C.1. Remaining anonymous

Being active on the internet and remaining anonymous is extremely difficult in western countries. The only safe way is through internet cafés. I do not encourage sending or receiving emails anonymously. The way it can be done, can be found on many websites.

7.C.1.A Sending anonymous emails

A pretty safe way of sending anonymous emails is using an internet café.

Whatever log files they keep, that information is probably not traceable to you anymore (unless you pay with a credit card or ID-papers were requested). Just create a new webmail account (*Hotmail*, *Yahoo!* etc.) each time you want to send emails and you are done.

7.C.2. Receiving emails anonymously

Open a free email account, like *Hotmail* or *Yahoo!*, from an internet café. And read the account only from other internet cafés. Every other way will leave traces. If the sender is traceable, your anonymity is lost as well.

I do not encourage receiving anonymous emails. This is just a warning to show how difficult it is to remain anonymous.

7.D Protecting your data

Nowadays you can backup you whole correspondence of the last 20 years on one USB disk or on a network drive. What would happen if somebody finds your disk and finds it interesting to read your old correspondence? Or even worse: what if the 'finder' has criminal actions in mind? How do you protect yourself from this hazard?

7.D.1. Office PC

I assume now that you are an administrator on your own PC. In the following I will suppose that you are on a *Windows* system. For other operating systems similar procedures exist.

7.D.1.A Protect partitions

Use a program like *Partition Magic* to make at least two partitions on your hard disk. The best is to have three partitions: a partition with drive letter c (*FAT*) with the installed programs of your company and of the computer supplier, a partition with drive letter d (*NTFS*) for your data and a partition with drive letter e (*NTFS*) for your own installed programs.

> Only partition d needs to be protected from the outside world. Change the access privileges to partition d by removing every user except SYSTEM and user yourself.

Now it will become quite difficult for other people to read your drive d: Your ICT people have to reset your password and log in pretending to be you, which is a criminal action for which they could be fired. You will discover that this illegal action has occurred because your password has become invalid. The downside to this safety precaution is that your data will not be backed up.

http://www.symantec.com/norton/products/overview.jsp?pcid=sp&pvid=pm80

7.D.1.B No traces

If you do not want other people to trace your activities on your computer I will give you some advice.

Remember that much of what you do on your computer will also be logged on other computers (like web servers, email servers, *NSA*, etc.). You do not have any control over this remote data.

To wipe your traces after internet surfing you should delete all your cookies. You should clear the history files of your browser. Browsers usually support this erasing feature.

> Empty recycle bins. Wipe/delete free space, special utilities (freeware) allow for this. Defragment your disk regularly.

Your employer may have installed spyware programs. Run anti-spyware software not supplied by your employer. Wipe/delete temporary files. There are many locations on your hard disks where temporary files are located. The *Norton CleanSweep* part of *Norton SystemWorks* can find them for you.

Your computer will be maintained by the ICT group. Unless you have taken special precautions, this means that they can read all the files on your computer.

http://www.symantec.com/norton/products/overview.jsp?pcid=sp&pvid=nswbasic2008

7.D.1.C USB disk

How to protect your USB stick or drive? I will give you the *Windows* solution. Format the USB disk as *NTFS*.

Make two root directories, for instance *pub* and *private*.

The idea is to make the folder pub readable by user *EVERYONE*. This directory can be used when you are connecting your USB-stick to a computer owned by somebody else, for instance to give a presentation.

I will assume you will use a USB-disk with a number of computers that are all under your control. Let us call them PC1, PC2, etc. I suppose you have a user account on all those PCs with allowance to set file-access rights. Connect the USB-disk to PC1 and add your user account and add *EVERYONE* to the group of users who are allowed to read the directory *private*. Disconnect the disk from PC1 and connect to PC2. Add your user account of PC2 to the allowed users of the disk. Etc.

Now the directory *private* will at least have the following users with full access rights: *EVERYONE*, UPC1, UPC2, and UPCetc, Probably user *SYSTEM* and user *Administrators* will also be there. Remove all users (EVERYONE, *Administrators*, etc.) except *SYSTEM* and your own user accounts.

People mounting the disk on another computer system will have quite some problem in getting at your data, as *NTFS* files are encrypted on the disk.

If you want to give a new computer access to the disk, you first have to add the user *EVERYONE* to the disk. Then mount the USB-device on the new computer. Add the appropriate user account and remove *EVERYONE*.

7.D.2. Network drive

Your network drive will be maintained by the ICT group. If you are lucky, they have set the access rights so that you can change these rights.

Request in any case that there will be a directory for which you have full rights. Limit access to this directory (and its subdirectories) to only yourself and the system. This does not exclude the ICT people from reading your files on the network drive and indeed reading your email. But it makes life more difficult for them.

As the network drive will be backed up regularly, all your email files – as a matter of fact all your files on the network drive – will remain in the universe forever.

7.D.2.A Encryption

You might want to encrypt old files. Encrypted files can probably still be read by the *NSA*, but normal beings like you and me would have to spend their whole life on decrypting it.

Encryption however, is only for people with very strong discipline. You should remember the password or otherwise the old emails are lost to you also. The drawback of encrypted files/directories is that they are in general not searchable any longer by a number of search programs.

Encrypted files that have already been backed up while they were still unencrypted, are not safe for people having access to those unencrypted archives.

8 INTERNET PROTOCOLS

The basic addresses on the internet are IP-addresses consisting of four octets separated by three dots: [0-255].[0-255].[0-255].[0-255]

As human beings are very bad in remembering numbers, people have introduced a way to translate IP-addresses into a string of characters.

For instance "www.stringcat.com" is an alias for IP-number "72.41.67.231", and IP-number "72.41.102.87" denotes the domain "www.secure.stringcat.com".

An additional advantage of using names is that the underlying IP-numbers are allowed to be changed without invoking the necessity of changing the names as well.

8.A Uri's

The location of files in the internet is denoted through url's (universal resource locators) and uri's (universal resource identifiers). The idea is that url's uniquely identify the location of a file somewhere in the world. In addition the locators give information on how to get at that information (for instance ftp, http, etc.). So a full uri could be "http://www.stringcat.com".

http://en.wikipedia.org/wiki/Uniform_Resource_Locator
http://en.wikipedia.org/wiki/Uniform_Resource_Identifier

8.B Name servers

The translation from names to IP-addresses is done through 'name servers', which are computers (or programs) dedicated to this task. There is a worldwide network of connected name servers that use extensive caching to minimize the necessary for look-ups. But finally a lookup request should end in a translation to an IP-number that the SMTP server of your provider, or your browser can use.

8.B.1. Spoofing

Double-lookup (translating the name into an IP-address and then vice-versa) is one of the techniques for spotting malicious users. Spoofing IP-addresses can be detected in this way. In addition, if your browser sends out an http request, both an IP-address and a server set name lookup can be used to detect credulous users.

8.C Core protocols

On the internet there are two core transport protocols: UDP and TCP.

Using UDP, programs on networked computers can send short messages known as datagrams to one another.

UDP can also stand for "Unreliable". This does not mean you will lose all your data, but it does not provide the reliability and ordering guarantees that TCP does. Datagrams may arrive out of order or go missing without notice. UDP does not have the overhead of checking if every packet actually arrived.

The TCP protocol guarantees reliable and in-order delivery of sender to receiver data.

http://en.wikipedia.org/wiki/User_Datagram_Protocol
http://en.wikipedia.org/wiki/Transmission_Control_Protocol
http://en.wikipedia.org/wiki/Packet

8.D Ports

A network port is a special number, ranging from 0-65535, recognized by the TCP and UDP protocols. These protocols use the ports to map incoming data to a particular process running on a computer. For TCP the following assignment holds for the ports:

0 to 1023: common,
1024 to 49511: registered,
49512 to 65535: private.

Examples of common ports: ftp 21, smtp 25, http 80, sol 443. Registered services are registered by third parties.

8.E Email RFC

The most important *Requests for Comments* regarding internet email are (SMTP) RFC 821, RFC822, RFC1521, RFC2047, (POP3) RFC1725, RFC1939, and (IMAP4) RFC1730, RFC2195.

ftp://ftp.rfc-editor.org/in-notes/rfc821.txt
ftp://ftp.rfc-editor.org/in-notes/rfc822.txt
ftp://ftp.rfc-editor.org/in-notes/rfc1521.txt
ftp://ftp.rfc-editor.org/in-notes/rfc2047.txt
ftp://ftp.rfc-editor.org/in-notes/rfc1725.txt
ftp://ftp.rfc-editor.org/in-notes/rfc1939.txt
ftp://ftp.rfc-editor.org/in-notes/rfc1730.txt
ftp://ftp.rfc-editor.org/in-notes/rfc2195.txt

9 ABOUT

9.A Abbreviations

Important abbreviations often used in the text are:

POP3	Post Office Protocol version 3
SMTP	Simple Mail Transfer Protocol
RFC	Request for Comments
IMAP4	Internet Message Access Protocol version 4

http://en.wikipedia.org/wiki/Post_Office_Protocol
http://en.wikipedia.org/wiki/Simple_Mail_Transfer_Protocol
http://en.wikipedia.org/wiki/Request_for_Comments
http://en.wikipedia.org/wiki/Internet_Message_Access_Protocol

9.B Trademarks

Acrobat and InDesign CS2, are registered trademarks of *Adobe Systems Inc.*
 http://www.adobe.com/
Arc Disk is a product of *Archos.*
 http://www.archos.com/?country=global&lang=en
Cisco vpn client is a product of *Cisco Systems, Inc.*
 http://www.cisco.com/
dtSearch is a registered trademark of *dtSearch Corporation.*
 http://www.dtsearch.com/
PayPal is a registered trademark by *eBay Inc.*
 http://www.ebay.com/
FileOpen WebPublisher is a registered trademark of *FileOpen Systems Inc.*
 http://www.fileopen.com/
GMail and Google Desktop are registered trademarks of *Google Inc.*
 http://www.google.com/corporate/index.html
The *SpamCop website* is maintained by *IronPort Systems, Inc.*
 http://www.spamcop.net/
DuplicateEmailRemover is a product of *MAPILab Ltd.*
 http://www.mapilab.com/

Exchange Server, Hotmail, Internet Explorer, Outlook and Windows are registered trademarks of *Microsoft Corporation.*
http://www.microsoft.com/en/us/default.aspx
Pegasus Mail was developed by David Harris. It is a free product.
http://www.pmail.com/
Eudora is a registered trademark of *Qualcomm Inc.*
http://www.qualcomm.com/
Beyond Compare is a registered trademark of *Scooter Software Inc.*
http://www.scootersoftware.com/
SPAMFighter is a product of *SPAMFighter ApS.*
http://www.spamfighter.com/
Norton CleanSweep, *Norton SystemWorks* and PartitionMagic are registered trademarks of *Symantec Corporation.*
http://www.symantec.com/index.jsp
Yahoo! is a registered trademark of *Yahoo! Inc.*
http://www.yahoo.com/

INDEX

abbreviations 75-76, 78, 80, 101, 114, 149, 249
absolute numbers 68, 160, 169
absolute statements 73
abstract 41, 43-44, 61, 74, 99, 102
– subjects 74
abuse of same symbol 80
accent 78, 133-134, 212
accepted for publication 65
access protection 112
accessories 181
acknowledge 36, 46, 63, 93, 97, 100, 115, 138, 170-171, 190, 208, 250
acknowledgement 45, 63, 93, 97, 100, 115, 138, 171, 190, 250
acquaintance 129
Acrobat 39, 83, 108, 112-114, 121, 124, 161, 165, 188, 190, 195, 249
– Distiller 83, 165
ACS 114
action box 173-174
active or passive form 74
actor 74, 133
adapter 181, 188-189
additional references 98
address books 219
administration 103, 211
Adobe Reader 112, 114, 121, 190
affiliations 45, 60-61, 170, 223
after care 186
afterburner 171
AIP 37, 114
alias 79, 203, 207, 211, 219, 229, 247
aliases 202, 215
align left 155
alignment 52, 54-55, 90, 148, 155, 159-160, 225
allotted time 133, 137
alphabet 77
alternative publishing 107
alternative referees 101
ambiguous 71-73, 222
Andale Mono 215
animated gif 172
animated text 172
animation 125, 152, 164, 171-174
anonymous 243
answering reports 95
Apple 50, 114, 121, 190
APS 37-38, 50, 61, 65-66, 114, 250
arbitrary units 68, 89
archive 65, 104, 107, 111, 165, 201, 237, 239, 243, 246
archived plots 165
archiving 39, 105, 230, 237, 239
arrows 77, 86, 158-159
Art HTML 110, 115
ASCII 43, 50, 84, 95, 101, 199, 204, 207, 211-214, 219, 221, 226-227, 238
assertive 126, 129, 214
speakers 126
at home 125, 174, 179, 186-187, 233-235, 241
atmosphere 122, 127, 143
attachment 95, 210-213, 216-217, 219, 221-222
attitude 130
authors 36, 39, 43, 45-47, 56, 58-60, 64-67, 73, 79, 88, 92, 95-96, 98-99, 101-102, 149

auto recovery 175-176
automated replies 227
automatic forwarding 229
axes 68, 82-83, 85-86, 145, 164, 166

background color 164
background picture 151
background texture 151
backup 142, 175, 183, 188, 232-234, 238, 241, 243
bad characters 215
balls 157-158
Base64 208, 214
BBC World Service 38, 124
bilateral correction process 45
bitmap 84-85, 108, 146, 161-164
blind copy 216, 220, 227
blockbuster 64
Bluetooth 180-181, 188
boards 186, 220
body 130, 132, 210, 215-219, 221-226, 228
– movements 130, 132
– position 131
bold 73, 77, 153-154, 160-161, 163, 212-213
bounding box 84
boxing equations 163
bullets 157-159, 169
business card 221-222, 229
busy slide 147-148, 153, 167, 170
busy figures 86

C++ 146
capitals 60, 154
caption 41, 57, 67, 82-84, 90, 102, 147, 165
CD-ROM Emulator 73, 115
cell phone 181-182, 188
chairman 126, 129, 132-133, 137-138, 140-141, 170, 186, 221
checking coauthors 46
checklist 188
choice of references 64
Cisco 233, 249
clause 57, 72, 80
cleanviews 206
clipping 147-148, 160
closing remarks 138
clutter 85, 148
CNN 38, 124
coauthors 36, 45-48, 54, 58-60, 62-63, 65, 75, 84-85, 95, 103-104
collection 30, 36, 87, 121-122, 136, 154, 157, 164, 194, 240
colloquium 120, 125, 135-136, 138
colons 158-159, 166
color 40, 54, 86-88, 124, 129, 150-152, 154, 156-158, 163-164, 169, 186, 196, 221
– coding 88, 154, 158, 169
colored background 150
colorful figures 87
coloring equations 162-163
COM 166
commandments 124, 187
comments 42-43, 46-48, 55, 71, 93, 96, 99, 102, 123, 132, 141, 143, 198-199, 202, 213, 248-249
commercial copy protection 112
commercial product 125
commercial reasons 111
communication with first author 47
compatibility 51, 84, 175, 213
composition 147-148, 155, 225
computer professional 242
computer programs 41-42, 124, 152, 176
conclusion 35, 57, 62-63, 68, 102, 145, 170-171
– slide 170-171
conference 41, 43-44, 126-128, 131-132, 135, 138, 144, 167, 175, 179, 181-188, 220, 2297-228
– abstracts 43-44
confidential 130, 216, 219
confirm by email 199
conflicting referee reports 98

consistency 41-42, 76
control any damage 216
contrast 86, 150-152, 154, 156, 161-164, 167, 187
conventional abbreviations 76
conventions 39, 41-42, 77-78, 124, 195, 215
copy right and web posting 110
copyright 110, 159
Cornell University 65
corrections by coauthors 45
corrections by one coauthor 46
corresponding author 45, 60, 95
count-down timer 133, 182, 188-189
Courier 94, 160, 215
coworkers 42, 170, 209
credit 39, 61, 64, 71, 105, 124, 138-139, 142, 144, 149-150, 170, 195, 211, 228, 243
criminal 211, 227-228, 243-244
criticizing prior work 71
cross-sections 88
culture 39, 111, 131, 224
cursor 225

dangerous colors 87, 152
database 51, 60-61, 66, 103, 107, 203, 219
datagrams 248
date 39, 45, 48, 53, 62, 104, 167, 187, 206, 217-218, 224-225, 240
– format 224
dead hyperlink 214
deadline 48, 101, 208-210, 216, 226
– for active corrector 48
default 52, 78, 82-83, 86, 142, 148, 152, 158, 160, 164, 176, 207, 211, 221-221, 231
– colors 152, 164
definite articles 58, 154-155, 158, 167
delay 84, 99, 129, 133-134, 183, 198, 210, 227
desktop publishing 110
diacritics 77-78, 212
diagonally browsing 35
dictionaries 53
digital archive 104, 107, 111
digital version 36-37, 47, 84, 122, 164, 195
discipline 34, 48, 105, 119, 137, 147, 168, 194, 232, 234-235, 239, 245
distrust 220
disturbing expert 141
Document Map 53
double quotes 40, 124, 195
drawing elements 85
Dreamweaver 110, 115
dressing 131
dtSearch 240-241, 249
duplicate emails 235
duplication 48-49, 150, 230
– problems 48-49

email address 202-203, 205-207, 210, 219-220, 224, 229
email client 195, 199-202, 204-207, 210-215, 219, 221, 223-227, 229-230, 232-237, 239-240
email server 199-201, 213, 243-244
email to all 204, 220
emphasize own work 72
emphasized and underlined text 73
encryption 112, 242, 245
end of you message 225
endless 210
Equation Editor 52, 161
equation signs and definition sign 79
error bars 67, 69, 86, 160
ESMTP 200
et al. 66-67, 101, 149
ethics 39
evaluation committee 129, 143
example of bad caption 83
Exchange Server 200-201, 219, 231-232, 235-236, 250

exclamation marks 73, 171, 204
exclamation sign 155
experiment 57-58, 61, 67-71, 83, 88, 102, 152, 166, 177

face to face 197
false colors 120
female speakers 134
figure 35-36, 41, 44, 51-52, 57, 67-68, 70, 75, 76, 82-87, 102, 104-105, 108-112, 119, 136, 145-148, 151-152, 155, 161, 163-166, 171, 176, 213-214, 219, 223-224, 229-230, 236
- captions 67, 83

FileOpen 31, 36, 42-44, 112, 115, 249
- *WebPublisher* 112, 115, 249

files 50-51, 53, 84-85, 95, 99, 104, 107-112, 121, 146, 150, 161, 164-165, 172-173, 175-178, 181, 183-184, 186, 188, 198-199, 202, 211-213, 221, 223, 232-234, 237-239, 243-247
filing system 219, 241-243
filter 110, 161, 164, 204-205, 207, 230-231, 235, 240
final format 47
first author 36, 44-45, 47, 58-60, 63, 66, 95
first discovery 111, 138
first-time claim 71
fixed pitch 101, 160, 215
fixed-line phone 181
Fixedsys 215
flip-charts 186
fluorescent colors 152
flying text 171
FOM-Institute 31, 115, 190, 250
font 40, 55, 73, 78, 82-83, 95, 101, 108, 147, 150-160, 162-166, 168, 171, 176-178, 183-185, 187-188, 207, 212-213, 215, 221
- color 40, 154
- saving problems 178
- size 55, 82-83, 147, 152, 155-160, 162, 164, 166, 168, 171, 212

footer 57, 170
foreign language 66
formal letter 203, 217
format 36, 44, 47, 84, 87, 101, 106-110, 112, 122, 147, 156, 159-160, 163, 172, 175, 194, 205, 207, 211, 213-214, 224-226, 237-238, 245
forwarding 211, 228-230
frames 85-86, 156-157, 165, 167, 172
fraud 207
front slide 168
full capital abbreviations 75
full justification 52, 55-56, 155
full screen 121, 161, 172
funnel 137
further study 37

Garamond 153, 213
generic scientific texts 41-43
good text frames 156
Google 66, 115, 149, 150, 158, 182, 190, 195, 201, 206, 229, 232-233, 237, 240-241, 249
graded background 151
grammar 72, 154, 223
graphics 94, 121, 152, 158, 163, 173
grid lines 86, 206
group event 224-225
group file server 104
GUI 51, 200, 206
guides 30, 34, 36, 78, 119, 122, 148, 155, 193, 195

habits 124, 134-135
hacker 111-113, 144, 233
handouts 129, 131, 144, 149, 178, 182-183, 189

hard copy 46-47, 62, 101, 144
hard returns 50, 55, 58, 159, 207, 214, 226
header 48, 57, 91, 96, 156-157, 170, 204-207, 209-210, 212, 215-216, 218-220, 222-223, 228, 230, 237
headings with math 81
hiring committees 129, 221
hoax 202, 207
horizontal alignment 159
hostile audience 143
hostile questions 143
hostile referee 64, 92-93
hotel 126-128, 181-182, 184, 186
Hotmail 229, 243, 250
html 107, 109-110, 113-115, 144, 161, 205, 207, 211-214, 224
HTML Protector 113-114
hum 186
humanists 38, 123
humor 96, 142, 222
hyperlinks 109-110, 113, 150, 196, 205, 211, 214, 216
hyperref 45, 109
hyphenation 52, 55
hyphens 52, 75

ibid 67
ICT 233, 242, 244-245
IEEE 37, 114
illegible 86, 165
IMAP4 200-201, 248-249
improving your English 38, 124
inbox 200, 202, 204, 208, 228, 235
  – copy 235
incremental texts 48
InDesign 55, 110, 114, 190, 249
index 61, 70, 116, 191, 240, 251
individual contributions 59
inline math 78, 80, 162
institutions in acknowledgement 63
integration with text 80
international bodies 78
internet 50, 61 , 65, 94, 104, 144, 172, 180-182, 184, 194, 198-202, 205, 211-213, 215, 224, 231, 233-234, 243-244, 247-250
interpretation 69, 88, 197
introduction 34, 57, 62-63, 137, 149, 168, 193
IP-address 247
IP-range 233
ISI Web of Science 115
italic 39, 73, 78, 124, 195, 212-213
IUPAC 78, 114

J11 jack 181
Java applet 172, 183
Junior Guide 34, 119, 183, 193
justification 51-52, 55-56, 90, 155

keywords 61, 66

labels 82-83, 85, 165-166
landscape 87, 147, 160, 164, 171
language 37-38, 42, 52, 58, 66, 70, 72, 75, 123-124, 129, 134-135, 146, 194, 215, 223
laptop 143, 146, 152, 168, 173-175, 179-186, 188, 201, 232
  – rebooting 179
  – resolution 180
laser pointer 180, 182, 188
LaTex 37, 39, 44-45, 50-52, 62, 84, 99, 107-109, 124, 161
left justification 55-56
legal disclaimer 125, 196
length of sentences 57
letter simulation 217
letter to the editor 92, 98-99
life internet 144
lighting 87-88, 160-170, 183, 185
limitations 39
line spacing 67, 147, 158-159, 162
line thickness 82, 86, 164
list of figure captions 67
list of references 43, 63-64, 66, 93, 98
local expert 138

locators 247
logos 163, 167, 187

Macintosh 50, 114, 121, 190
macros 50, 53-54, 161, 167
mailing list 103-104
male chauvinism 39, 123, 194
malicious 210, 247
manager 30, 44, 48, 54, 61, 104, 106, 170, 197-198, 205-206, 209-210, 216-217, 225, 242
managing email accounts 231
manual 37-39, 41, 43, 52, 55, 108-109, 159, 166, 177, 180, 231, 236
manuscript handling 41
manuscript modifications 97
manuscript types 41
margins 46, 48, 52, 55, 83, 146-148, 187
master slide 146-148, 156, 166-168, 187
math 44, 62, 77-78, 80-81, 94, 160-163
Mathematica 88, 155
MathType 44, 53, 114, 161, 188, 190
meeting of authors 47
Merriam Webster 73, 115
message header 206-207, 215, 219, 228, 230, 237
message to experimentalist 70
message to theoreticians 69
missing fonts 176
mobile phone 180-181
mode of transport 94
monospace 207, 215
movable storage 234
MS-Office 52, 161
MS-Office 2003 51, 53, 175
MS-Word 37, 45, 50-54, 56, 85, 99, 109-110, 112, 115, 164, 198, 211, 221
multi-lined texts 155
multiple accounts 229-230, 235-236
multiple POP3/SMTP servers 230

name servers 247
naming conventions 78
nature 88, 93, 99-101, 107, 110, 135
Nature and *Science* 88, 93, 99, 101, 107, 110, 137
navigation 35, 37, 45, 53, 79, 109-110, 113, 119, 122, 173, 195
nervousness 134
network drive 201, 232-235, 241, 243, 245
New York Review of Books 38, 124
NIST 78, 114
non-alphabetic characters 76
non-commercial spam 205
non-journal references 66
non-postscript 85
non-serif 153
novel 71, 154
NSA 244-245
numbered outline 158
numbering 53, 56-57, 71, 79, 158, 162, 168-169, 187
– of equations 79

obligatory items 58
obligatory slides 166
old *PowerPoint* version 177
one referee at a time 97
one story 137
one-liners 120, 171
online 44, 65, 73, 109, 204
opening remarks 96, 138
– rebuttal 96
operator 61, 137
opt-out 206
order of authors 59
organization of content 55
Origin 83, 87, 115, 164, 188, 190
OSA 37, 114
other text formatters 51
outlined text 44, 157-158, 172
Outlook 137, 157, 171, 178, 195, 199-200, 206, 209, 211, 213, 216, 219, 225-226, 231-241, 250

out-of-office reply 227
own work 44, 64-65, 70, 72, 96, 110, 139, 149

pace 131, 133-134
Pack-and-Go 177
PaintShop 163
paragraphs 37, 43, 56, 97
parking places 57
partitions 246
patchwork 152, 156, 164
PCMCIA 180, 188
pdf files 107-109, 112, 121, 150, 198, 221
people in acknowledgement 63
periods 154, 158
Perl 200, 211
permission 92
perseverance 98
phone number 182, 223
PhotoPaint 164
photos 163
PhotoShop 161, 163-164
pictures 89, 94, 108, 113, 146, 150, 158, 163-164, 213, 231
plugs 181, 184
pointer 128, 180, 182, 188
POP3 200-201, 205, 229, 231-232, 234-235, 248-249
popular talks 136
port 181, 248
postscript 84-86, 108-109, 163
power management 179, 184
PowerPoint 84-86, 115, 121-123, 147, 150, 155-159, 161, 167, 171-178, 184-185, 188, 190
preference schedule 127
preparation time 125
preprint server 65, 107
prestige 229
pride 100
printed version 36-37, 40, 48, 55, 122, 124, 195-196
printing problems 131, 178
prior knowledge 37, 129
priority 66, 70, 107, 138, 142, 166, 236
 – claim 138, 142
privacy 242-243
private communication 65-66, 93
professional examples 159
professional lighting 88
progress 41, 48, 95, 165, 167-170
proportionality sign 68
proprietary fonts 83, 176
protecting 111-112, 243
– pdf files 112
– solutions 111
– your papers 111
protocols 194, 198-200, 215, 247-248
pst files 234, 238
publication costs 103
publication form 36, 122, 195
publicizing your work 106
punctuation marks 79, 166

quality of manuscript 102
question 45, 59-60, 97, 99, 120, 126-129, 133-134, 137-143, 155, 186, 242
 – mark 155
quick response 95, 208

RAID 233
raw email address 202
recap 169, 217
recital 133
record 50, 60, 93, 239
recovery 52, 175-176
redirecting 228
referee finds a flaw 97
referee reports 41, 44, 95, 98-100, 104
references 35, 43-45, 61, 63-64, 66, 71, 93, 98-99, 119, 149, 150, 171
refused coauthors 63
regular publications 111
rehearse 128-129, 131-134, 143, 179, 187

relative dates 224
repair figures 166
repetitive email 204
reply address 229, 235
reply to all 227
reprint orders 103
Requests for Comments 198-199, 248
required prior knowledge 37
response forms 210
response time 216
retrieve 235, 237, 239-240
retrieving emails 235
resubmission package 99
reusability 146-147, 167, 176
reuse 36, 84, 86, 91, 111, 125, 148, 169, 183, 219
reviewers 43, 66, 99, 101, 132
reviewing load 101
room mates 128
rotating figures 171

salutation 217, 220, 222-223
same words over and over again 72
sans serif fonts 153
saving files 104
scanned images 104, 198
science agencies 41, 170
scientific conflicts 97
scientific level 135
scientific plots 163-164
Scientific Word 50, 115, 161, 190
secretaries 194, 198, 219, 221, 225, 242
security 144, 194, 202, 229, 233, 242
self-citations 64
self-contained 34, 37, 90, 119, 122, 193, 195
seminar 121, 126, 135-136, 138
Senior guide 34, 119, 128, 193
sensitive emails 216
sent emails 193, 201, 239
serif fonts 153
service pack 178
shake hands 129
signature file 222-223, 226
single quotes 40, 124, 195
size constraint 37, 42, 69, 123, 194
size limit 47, 64
slick presentations 125
slide background 151
slow servers 233
smileys 158
smooth text transitions 57
SMTP 200-201, 203, 230-232, 236, 247, 248-249
snowing transitions 171
social behavior 138
social duty 100
sockets 181, 228
soft forward 230
soft returns 214
software incompatibility 174
some special scientific texts 42
sound 130, 186
source text 110
spaghetti text 56, 184
spam 202, 204-207, 235
SpamCop 206, 249
SPAMFighter 205, 250
speech support 134
speech therapists 134-135
spelling 60, 66, 75-76, 99, 119, 154, 170, 198, 222-223
– errors 75, 99, 198, 223
– of names 60, 66, 170
spoken text 133
SSL 228
stall 208, 227
standard format 84-85, 101, 163
standard partitioning 57
standard reply 226
stopgaps 135
StringCat 108, 247
students 30-31, 34, 115, 119, 128, 136, 173, 190, 194, 198, 227, 242, 250
style manual 37-38
subheadings 53, 56

submission 34, 44-45, 50, 62, 92-94, 99-100, 103-104
  – letter 92, 94, 100, 104
surveys 210
symbols in equations 80
synchronization 48, 185, 232-234

tables 61, 70, 90, 102, 119, 160, 206
talk 120-122, 125-141, 143-144, 149-150, 156, 158, 168-172, 178-179, 181-183, 185-186, 188, 215-216
target group 34, 37, 119, 123, 194, 206
TCP 242, 248
technical text 194
telegram style 154
template 53, 147, 167
ten commandments 184
Tex 37, 50, 52, 55, 107, 110, 161
TexPoint 161, 188
text aspects of figures 82
text boxes 156
text content 58, 113
text formatter 37, 50-51, 78
text frames 156
text properties 153
text spelling 75
text structure 36, 55, 225
theory 58, 68-70, 79, 82, 88, 102, 107, 140
  – and experiment 68-69, 82
Thesaurus 53, 72-73, 115
tick marks 82, 85-86
timer 133, 182, 188-189
title 48, 57-58, 63, 90, 96, 99, 101, 103, 142, 147, 155-157, 165-168, 170-171, 174, 217-218, 223
  – slide 168, 170
TMI 136, 223
to-do list 125, 197, 200, 202, 208-209, 216, 227, 230, 235
trademarks 114-115, 190, 249-250
training courses 123
transition 57, 132, 148, 171-172, 174
transparent 75, 112, 164, 219, 228, 230
trap 101, 142-143
trimming 226
trivial body 223
True Type 108
Type 1 83, 108
Type 3 108, 150

UDP 242, 248
underline 62, 73, 196, 212
underlining 154, 156
units and constants 78
University of Amsterdam 31, 115, 190, 250
University of Twente 31, 115, 190, 250
Unix 50, 84, 121, 201, 215, 221, 239
unofficial publishing 107
unpublished 65, 144
  – data 144
  – material 65, 144
uri's 247
url's 247
USB device 184, 234, 241, 245
use of appendices 80
use of first names in author list 60

vector format 84, 163
vectors 77, 161
Verdana 153, 213
version control 47, 54, 240
vertical alignment 159
video 173
video projector 86, 129, 151-152, 173, 179-180, 184-185
VPN 165, 230-231, 233, 249

waiting time 48, 227
wall plugs 181
watermarks 111-112, 144, 183
Webdings 158

webmail 200-201, 232, 243
website 38, 92, 106, 110-111, 113, 123, 144, 149-150, 249
where to publish? 45
white space in concepts 46
widows 54
Wifi 180, 184, 188
Window 172-173, 206, 214, 225
Windows 50-51, 84, 108, 115, 165, 173, 176, 190, 201, 221, 239, 243, 245, 250
 – *Media Player* 173
Winedt 50, 114
wireless mouse 180, 188
women 39, 123, 194
wrap 67, 112, 158
writing referee reports 100

X-axis 85, 87, 165

Yahoo! 229, 243, 250
Y-axis 68, 85, 87, 165